Studies in Smart Technologies

Advanced communication and computing technologies to develop intelligent and adaptive systems are usually referred to as smart technologies. Many of the sustainable development goals (e.g., SDG 4, SDG 6, SDG 7, SDG 8, SDG 9, SDG 11) cannot be achieved without developing smart technologies. The aim of the proposed book series, "Studies in Smart Technologies," is to provide a detailed and latest overview of the latest research findings in the field of smart technologies and allied areas. The series covers studies related to smart cities, IoT, smart sensors, AI, soft computing, and other related areas. The content published in the series will be an important reference to the policymakers, industries, researchers, practitioners, and students in computer science, electrical engineering, information technology, and aerospace engineering.

The series will publish monographs, edited volumes, textbooks and proceedings of important conferences, symposia, and meetings in the field of smart technologies and their applications.

Venkata A. P. Chavali · Amit A. Deshmukh ·
Aarti G. Ambekar

Neural Network Modelling of Microstrip Antennas

Springer

Venkata A. P. Chavali
Department of Electronics
and Communication Engineering
Indian Institute of Information Technology
Surat, Gujarat, India

Aarti G. Ambekar
Department of Electronics
and Telecommunication Engineering
SVKM's Dwarkadas J. Sanghvi College
of Engineering
Bhakti Vedanta Swamy Marg
Mumbai, Maharashtra, India

Amit A. Deshmukh
Department of Electronics
and Telecommunication Engineering
SVKM's Dwarkadas J. Sanghvi College
of Engineering
Bhakti Vedanta Swamy Marg
Mumbai, Maharashtra, India

ISSN 3004-9881 ISSN 3004-989X (electronic)
Studies in Smart Technologies
ISBN 978-981-96-9430-3 ISBN 978-981-96-9431-0 (eBook)
https://doi.org/10.1007/978-981-96-9431-0

This Springer imprint is published by the registered company Springer Nature Singapore Pte Ltd.
The registered company address is: 152 Beach Road, #21-01/04 Gateway East, Singapore 189721, Singapore

Preface

The rapid convergence of Machine Learning with Electromagnetics and Antenna Engineering has opened unprecedented avenues for innovation, design optimization, and performance enhancement. This textbook, *Neural Network Modelling of Microstrip Antennas*, aims to provide a comprehensive introduction and in-depth exploration of how machine learning based Neural Network Techniques are reshaping antenna research and development.

The motivation for this work stems from a clear gap in the available literature; while numerous resources cover traditional antenna theory or general machine learning, few address the intersection where these fields meet. This book is designed to bridge that gap by offering both introductory knowledge and advanced insights into applying machine learning, and optimization algorithms to solve complex antenna design challenges.

The material presented here is the result of years of research, experimentation, and collaboration with students. Throughout, the focus has been on providing clear explanations, and practical examples, that demonstrate how Neural Network can complement classical Microstrip Antenna Design methods to achieve superior results.

This book is intended for graduate students, researchers, and practicing engineers who wish to understand and apply Neural Network Models in the context of Microstrip Antenna modelling, analysis, and design. Readers are expected to have a basic foundation in Electromagnetic Theory and an interest in exploring the computational and data-driven methods that are now shaping the future of this field.

Any errors or omissions remain our own responsibility, and we welcome readers' feedback and suggestions for improvement in the future editions.

It is our hope that this book not only informs but also inspires further innovation at the intersection of Machine Learning and Microstrip Antenna Technology.

Surat, India — Dr. Venkata A. P. Chavali
Mumbai, India — Prof. Amit A. Deshmukh
Mumbai, India — Dr. Aarti G. Ambekar
May 2025

Competing Interests The authors have no competing interests to declare that are relevant to the content of this manuscript.

Contents

About the Authors

Venkata A. P. Chavali received B.E. degree in Electronics and Communication Engineering from Andhra University, Vizag, India in 2004 and M.E. degree in Digital Communication from RGPV, Bhopal, India in 2010. She has completed her Ph.D. in Electronics and Telecommunication Engineering from D. J. Sanghvi College of Engineering under Mumbai University, Mumbai, India. She is currently working as an Assistant Professor (Level-12) with the Department of Electronics and Communication Engineering, at Indian Institute of Information Technology, Surat. Earlier she worked as Assistant Professor at Department of Electronics and Telecommunication Engineering, at SVKM's D. J. Sanghvi College of Engineering, Mumbai. Her research interests include Broadband Microstrip Antennas. To her credit she has 30 Journal and 54 conference publications, and two granted patents. She is on the reviewer list of various reputed journals such as *IEEE Transaction of Antennas and Propagation, International Journal of RF and Microwave Computer-Aided Engineering, Elsevier Micro and Nanostructures*, and various international conferences.

Amit A. Deshmukh obtained B.E. (Electronics) from V.I.T., Pune University in 1997. He obtained M.Tech. Degree in 2000 and Ph.D. in 2004 from Department of Electrical Engineering, IIT Bombay. His thesis work during masters and doctoral, was on Compact broadband and dual band microstrip antennas. He worked as Research Assistant in Department of Electrical Engineering, IIT Bombay. Further, He worked as Assistant Professor in Sardar Patel Institute of Technology, Mukesh Patel School of Technology Management and Engineering (NMIMS–DU) and D. J. Sanghvi College of Engineering. He also worked as Member of Technical staff in R&D division of Air Tight Network Pvt. Ltd., Pune. Currently he is working as Professor and Head of the Department of Electronics and Telecommunication Engineering at SVKM's D. J. Sanghvi College of Engineering, Vile—Parle (W), Mumbai, India. He has together (Teaching, Industry, and Research) more than 22 years of experience. He has published more than 400 research papers in various International and National Journals and Conferences, and has two granted and two published patents. He is on reviewer list of many International Journals like, *IEEE Transaction of Antennas and*

Propagation, IET Microwave Antennas and Propagation, PIERS Journals, International Journal of Electronics, IETE Journal of Research, Elsevier Electronics and Communication Journal, etc. He has delivered many lecture talks in National/ International conferences and workshops in the areas of patch antennas.

Aarti G. Ambekar received B.E. degree in Electronics and Telecommunication Engineering from Cummins College of Engineering for Women, Pune University, India in 2007, M.E. degree in Electronics from KJSCE, Mumbai University, India in 2013, and completed her Ph.D. in Electronics and Telecommunication Engineering from SVKM's D. J. Sanghvi College of Engineering, Mumbai University, India in 2022. Since 2015 she is working as an Assistant Professor with the Electronics and Telecommunication Engineering Department, D. J. Sanghvi College of Engineering, Mumbai, India. Her research interests include Microstrip Antennas, Image Processing, and Computer Vision. To her credit, she has a total of 11 Journal and 29 conference publications, and two granted patents.

Chapter 1
Artificial Neural Network

Abstract Artificial neural networks are computational models inspired by the structure and functionality of the human brain, designed to identify complex patterns and relationships within data. Comprising interconnected nodes or neurons, artificial neural networks process input data by applying weights and biases, followed by activation functions that determine whether a neuron 'fires' to influence subsequent layers. Through iterative training processes such as backpropagation and the adjustment of weights and biases over multiple epochs, artificial neural networks learn to minimize prediction errors and improve performance. The architecture of a neural network including the number of layers and neurons plays a critical role in its effectiveness, with improper configurations leading to overfitting or underfitting. The success of an artificial neural network also depends heavily on the quality and volume of the training data. A well-structured neural network, trained on clean and extensive data sets, can produce highly accurate predictions, making artificial neural networks a foundational tool in modern artificial intelligence and machine learning.

1.1 Introduction

Neural networks are a set of algorithms that aim to understand the underlying pattern and relationship in a set of data. It is a mathematical approach to mimic the way a human brain works. The etymology of neural network suggests that the word "neural" refers to the analogy between artificial neurons of the neural networks and the organic neurons of the human brain. Neural network is often called ANN, referring to these set of mathematical algorithms that attempt to mimic human neural system. A neural network consists of thousands and even millions of simple processing nodes that are densely interconnected. All these connections have two parameters associated with them. Each of these connections is assigned a number called weight. The weight is a mathematical representation of the contribution the node is doing towards the output. When a node is activated, when it receives a different data item, a different number, the incoming number is multiplied with the associated weight. It then adds a bias value to the product and gets a single value. If this value is lower than a certain

V. A. P. Chavali et al., *Neural Network Modelling of Microstrip Antennas*, Studies in Smart Technologies, https://doi.org/10.1007/978-981-96-9431-0_1

threshold, then this node will not send data to the next layer. If the value is above the threshold, then the node is said to be active or fire. In this instance, the node will send data to the next layer. When a neural network learns, the learning process changes these associated weights and that is the learning process of a neural network. Weights and biases are the characteristic of a neural network.

The most basic unit of an ANN is called a neuron or a node. The node has a weight and bias as the two properties. The smallest ANN can have one neuron and there is no limit to how many neurons can be added to a network. The input to a neuron is fed into an activation function after being multiplied by the weight and the bias added to the product. The intention of an activation function is to output a value that satisfies the intent of the network model. If the input and output of a node are related as expected, then the neuron is 'activated' or 'fired'. This is determined by an error calculated between the prediction and truth values. This error is used to update the weights and biases of the neurons. This process is known as backpropagation. This process is repeated multiple number of times to 'fit' the input data to the desired output data, such that the weights and biases are changed enough to a point where the change happening is either zero or minimum. The number of times this process occurs for a data set is known as epochs. This whole process is known as training. The selection of number of layers and number of neurons in these layers are an essential part in designing a neural network architecture. If the number of layers is too many then the problem of overfitting arises. When the model returns a good result on a training data set but performs badly on a test data set then this is defined as overfitting. If the number of layers is too less than the network might not learn, and this is termed as underfitting. One of the methods to overcome these shortcomings is to have a big data set. Having a big and clean data set is crucial to developing an accurate artificial neural network model. Data processing asserts the discipline of supervised learning and ensures efficient and highly accurate predictions.

1.2 ANN Activation Functions [1]

In order to provide ANNs, the ability to learn and represent complicated relationships within data, activation functions are used to bring nonlinearity into the network [1]. In ANNs, the following activation functions are frequently used [1].

1.2.1 Sigmoid Function (Logistic)

The function can be defined as

$$f(x) = \frac{1}{1 + e^{-x}} \tag{1.1}$$

which yields values in the range of 0–1.

Due to issues with fading gradients, this function, which produces values in the range of 0–1, is no longer commonly used in hidden layers.

1.2.2 Hyperbolic Tangent Function (Tanh)

The function can be defined as

$$f(x) = \frac{e^x - e^{-x}}{e^x + e^{-x}} \tag{1.2}$$

This function helps to better handle negative inputs by being similar to the sigmoid function but with a zero centre and a range of − 1 to 1.

1.2.3 Rectified Linear Units (ReLU)

ReLU is an activation function that is employed in ANNs. Because of its efficiency and simplicity, it's one of the activation functions that deep learning models employ the most.

The function can be defined as

$$f(x) = \max(0, x) \tag{1.3}$$

To put it another way, it returns the input value for any input that is more than zero and zero for any input that is less than zero. Its derivative is one in mathematics for positive input values and zero for negative input values.

Among the benefits of utilizing ReLU are the following.

Computational Efficiency: Compared to other activation functions like sigmoid or tanh, ReLU is computationally more economical since it just requires basic thresholding operations.

Sparse Activation: This technique reduces the chance of overfitting in some neural network topologies by helping to achieve sparsity by zeroing out negative values.

However, ReLU is not without issues. The 'dying ReLU' phenomenon is one such problem when neurons stop updating their weights during training and become unresponsive to input. In order to mitigate the 'dying ReLU' problem, many variations of ReLU, including LReLU, PReLU, ELU, and others, have been developed to address the drawbacks of the basic ReLU function, while maintaining its benefits.

1.2.4 Exponential Linear Unit (ELU)

Neural network activation functions include the ELU. ELUs can reduce mean unit activations to zero, like batch normalization, but with less computing cost than ReLUs, because they contain negative values. Because there is less bias shift effect as the mean changes towards zero, learning is accelerated by moving the normal gradient closer to the unit natural gradient. Although LReLUs and PReLUs also contain negative values, a noise-robust deactivation state is not guaranteed by them. Smaller inputs cause ELUs to saturate to a negative value, which reduces the amount of variation and information that is forward propagated.

This function can be defined as

$$f(x)=\begin{cases} x & \text{if } x > 0 \\ \alpha(e^x - 1) & \text{if } x \le 0 \end{cases} \tag{1.4}$$

The responses of above activation functions are shown in Fig. 1.1.

1.3 ANN Optimization Algorithms

Finding an objective function's set of inputs that produces either a maximum or minimal function evaluation is known as optimization. Fitting logistic regression models and training artificial neural networks are only two examples of the machine learning methods that are based on this difficult topic. Many optimization algorithms are widely used; in major scientific code libraries, there are probably hundreds of algorithms available. As a result, choosing the right algorithms to consider for a particular optimization issue may be difficult [3].

Following are the different types of algorithms used to train the ANN model.

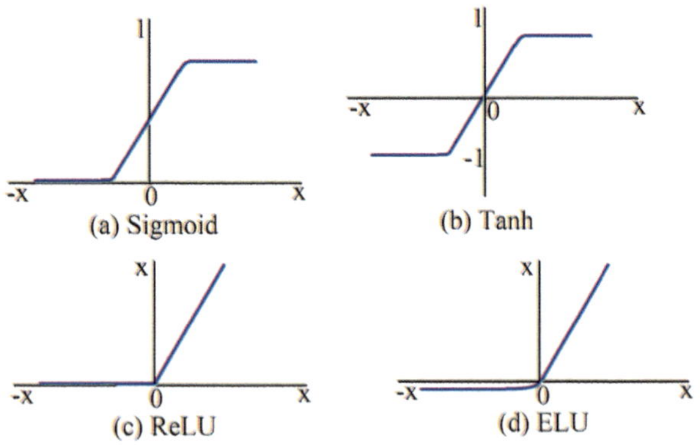

Fig. 1.1 ANN activation function [2]

1.3.1 Gradient Descent

The most fundamental but popular optimization approach is gradient descent. It plays a major role in algorithms for classification and linear regression. A gradient descent approach is also used in neural network backpropagation.

1.3.2 Stochastic Gradient Descent (SGD)

It is an adaptation of Gradient Descent. It attempts to adjust the model's parameters more often. This involves computing the loss on each training example and then adjusting the model parameters. Therefore, SGD, rather than Gradient Descent, updates the model parameters 1000 times in a single data set cycle if the data set has 1000 rows.

1.3.3 Mini-batch Gradient Descent

It is the most effective gradient descent algorithm variant available. Compared to normal gradient descent and SGD, it is an improvement. Following each batch, it modifies the model parameters. Thus, the data set is split up into different batches, with each batch having its own set of parameters updated.

1.3.4 Momentum

The creation of momentum softens the convergence and lowers the high variance in SGD. It decreases the fluctuation in the irrelevant direction and quickens the convergence towards the relevant direction.

1.3.5 Nesterov Accelerated Gradient (NAG)

Although momentum can be a useful tool, an algorithm may miss the local minima and keep increasing if the momentum is too high. Therefore, the NAG algorithm was created to address this problem.

1.3.6 Adagrad

The learning rate is constant for all parameters and for every cycle, which is one of the drawbacks of all the optimizers described. This optimizer modifies the rate of learning. Every parameter and time step (t) has a different learning rate (η). This algorithm for optimization is of the second order. It utilizes an error function's derivative to function.

1.3.7 AdaDelta

It is an expansion of AdaGrad that aims to fix its declining learning rate issue. As an alternative method gathering all previously squared gradients, Adadelta restricts the window of previously accumulated gradients to a predetermined size, w. Instead of using the total of all the gradients, an exponentially moving average is employed in this.

1.3.8 Adam

Momentums of the first and second order are handled by Adaptive Moment Estimation (Adam). Adam's theory is based on the idea that, in order to conduct a thorough search, we should gradually reduce our velocity rather than rolling over the minimum quickly. Apart from maintaining an exponentially decreasing mean of previous squared gradients, such as AdaDelta, Adam additionally maintains an exponentially decreasing mean of previous gradients. Adam is an excellent optimizer. Adam is the optimizer if someone wishes to train a neural network faster and more effectively. Consider using optimizers with a dynamic learning rate for sparse data. Min-batch gradient descent is the ideal choice if one wishes to employ the gradient descent method.

1.4 Summary

Over the past few years with the advent of AI, we have seen its use over a wide range of topics. Using ML, a subset of AI, a model based on past experiences (learning) and using it to predict the future can be developed. Microstrip antennas have risen in prominence in recent times due to their features like small dimensions, simplicity, and low cost of production. Neural networks can be used to establish relations between antenna parameters in ways that could not be done before. By carrying out simulations and feeding the network with large sets of training data, we can automate a tedious cog in the process of microstrip antenna design. A plethora of algorithms are available to us that can be used for establishing relations and predicting outputs for cases like Levenberg–Marquardt algorithm, Genetic algorithm, etc. These algorithms can be used to predict antenna parameters such as patch dimensions and slot dimensions with high accuracy leaving some room for potential optimization. Hence, they reduce a lot of the work done in arriving at parameter values through trial and error by automating that task. In the further chapters, artificial neural network models are used to predict antenna dimensions across a wide range of frequencies. This enables us to speed up antenna production by removing trial and error that goes into determining the ideal antenna configurations for a given application. Neural networks of two kinds are tested in this book. The first one involves only a single step between input parameters and antenna dimensions. The second kind involves two steps. With the help of these artificial neural networks, we can obtain dimensions in lesser time. The neural network is trained on actual simulation data and then tested against different frequencies. Through training by the Levenberg–Marquardt algorithm, we get satisfactory agreement between the results given by the neural network and simulations performed with CST Studio Suite [4], IE3D simulation software [5]. The measured data for fabricated antennas realized using prediction by neural network model also shows good accuracy in results against the simulation. For all the antennas, ANN codes are written using programming language 'Python' [6].

References

1. https://www.geeksforgeeks.org/activation-functions
2. https://medium.com/@shrutijadon/survey-on-activation-functions-for-deep-learning-9689331ba092
3. Doshi S (2019) Various optimization algorithms for training neural network. Towards Data Sci 13
4. CST Microwave Studio, Version 2019
5. IE3D simulation software, Version 12
6. http://www.python.org/

Chapter 2
Rectangular Slots-Loaded Microstrip Antenna for Broadband Response

Abstract This chapter presents the application of artificial neural network in the design of wideband and circularly polarized microstrip antennas based on variations of the E-shape configuration. Embedding resonant slots within air-suspended microstrip patches enhance bandwidth without increasing patch area, enabling compact single patch and high-performance antenna designs. The study focuses on regular and offset E-shape microstrip antennas, asymmetrical slot-loaded microstrip antennas, and multi-slot configurations for achieving broadband and dual band circular polarization across different frequency ranges. A low-cost FR4 substrate suspended above the ground plane is used in all the designs. Artificial neural network models are trained on extensive data sets covering substrate thicknesses from $0.02\lambda_g$ to $0.1\lambda_g$ and frequencies from 600 to 6000 MHz. Each dataset includes key antenna parameters such as patch and slot dimensions, resonance frequency, and feed point location. The trained artificial neural network models predict optimal design parameters with high accuracy, resulting in simulated resonance frequencies with errors below 2%. This approach facilitates efficient and accurate design of wideband and circularly polarized slot-loaded microstrip antennas over a broad range of operating conditions.

2.1 Introduction

One of the popular techniques to realize a single patch wideband, high-gain microstrip antenna is by embedding the resonant slots inside the air-suspended patch. This technique enhances the antenna BW using a single patch, and without increasing the patch area. Designs of RMSA loaded with a rectangular slots pair (E-shape MSA) are reported [1–8]. It is well known that in addition to reducing the higher order resonant mode frequency, the E-shape MSAs yield probe inductance compensation, which supports the increment in the BW. By incorporating multiple rectangular slots and their variations, various wideband designs of modified E-shape MSAs are reported [9–21]. Using additional slots, 5–10% increment in the BW of basic E-shape MSA is achieved. Utilizing E-shape MSA's symmetry across the

V. A. P. Chavali et al., *Neural Network Modelling of Microstrip Antennas*, Studies in Smart Technologies, https://doi.org/10.1007/978-981-96-9431-0_2

axis of feed point, its compact and wideband designs are reported [17, 18]. Using the resonant mode-based approach, a thorough study explaining the functioning of wideband behaviour in E-shape MSA and subsequent design formulation is reported [12, 21, 22]. A detailed comparison among the various wideband modified designs obtained using basic E-shape configurations, on the antenna output parameters like BW, gain, and radiation pattern, is reported in [19]. By incorporating asymmetrical slots (unequal in lengths) [23] and offset feed location in MSA, dual polarized or circular polarized response is obtained [20, 24–28]. In most of the designs offering wideband, multi-band response, electrically thicker substrates are used since they achieve higher reflection coefficient (S_{11}) BW and radiation efficiency [1–5]. Thus, in slot cut patch antenna designs, substrate thickness more than $0.06\lambda_g$ is selected. While realizing regular shape MSAs on electrically thinner substrates, design equations for calculating the patch dimensions at the fundamental mode frequency are mentioned in the literature [1–5]. But when these equations are used for electrically thicker substrate, then error with respect to simulated and measured results occurs. The error is attributed to the calculation of edge extension length in the dimensions of microstrip patch because of the presence of fringing fields. Accurate knowledge of this extension length is essential for calculating the patch dimensions on thicker substrate at the fundamental mode frequency. In E-shape design, wideband or circular polarized response is present around the fundamental TM_{10} mode. In the reported literature, design methodology for E-shape variations is presented for a limited range of substrate thicknesses and frequencies. The same is missing over a wide range to evaluate the antenna parameters like slot position, slot length, and the dimensions of basic MSA in wideband and circular polarized variations of E-shape MSAs. This requirement leads to the application of the ANN model in the designs of variations of E-shape MSAs.

This chapter presents an ANN model for air-suspended variations of rectangular slots cut MSAs such as regular E-shape MSA for broadband response, offset E-shape MSA for broadband and CP response in different frequency spectrums, asymmetrical position of slot-loaded RMSA for CP response, and multiple slot cut RMSA with offset feed for broadband and dual band CP response. In the case of suspended dielectric substrate, low-cost lossy FR4 substrate ($\varepsilon_r = 4.3$, $h = 0.16$ cm, $\tan\delta = 0.02$) located above the ground plane using finite air gap (h_a) has been used. Initially, the functioning of each antenna with respect to various resonant modes leading to broadband and/or CP response is explained in detail followed by its ANN model. In the ANN modelling for each configuration, each neural network is trained using various data sets of a large frequency range and substrate thicknesses. These data sets include various parameters of antennas such as thickness of the substrate, resonance frequency as a function of operating wavelength, patch dimensions such as length, width, slot length, slot position, and coaxial feed point location obtained using simulation for the given resonance frequency, and the half-wavelength value at the considered frequency. The data sets for training ANN are included at the frequency intervals of 200 MHz over 600–6000 MHz frequency band. The ANN code is written using programming language ‘Python’. Next, utilizing the trained ANN model, dimensions of E-shape MSA are predicted over thickness of the substrate in the range of 0.02–0.1λ_g and across 600–6000 MHz, frequency spectrum. All MSAs when simulated using these predicted dimensions using [29, 30], give approximately the same resonance frequency as the desired value showing percentage error less than 2%. All dimensions in the text and tables throughout the chapter are specified in cm, and frequencies are in MHz.

2.2 ANN Models for Slot Cut RMSAs for Broadband and CP Response

2.2.1 ANN Model for E-Shape Microstrip Antenna

In slot cut configurations, BW enhancement takes place due to the space tuning between the fundamental and higher order mode resonance frequencies. This is achieved by cutting slots at the maximum current location of the higher order modes. One such configuration is an E-shape MSA. The E-shape MSA consists of a pair of rectangular slot cut RMSAs as shown in Fig. 2.1a, b. To understand the functioning of E-shape MSA, the design reported in [19] is considered. Here, initially the basic RMSA is designed for fundamental TM_{10} mode frequency of 900 MHz. This frequency is selected as it covers the GSM band. For this frequency, the dimensions of the RMSA are identified as $L = 13$ and $W = 15$ cm.

The MSA is designed on suspended FR4 substrate with air thickness $h_a = 2$ cm and is fed using a 50 Ω coaxial N-type connector. Resonant modes of the RMSA are identified by studying the surface current distributions as shown in Fig. 2.1c–e. The E-shape MSA is optimized for BW by introducing a pair of rectangular slots of dimension '$l_s \times w_s$' as shown in Fig. 2.1a. These slots are orthogonal to the surface currents at TM_{02}, TM_{12}, and other higher order mode frequencies. An increment in slot lengths reduces their frequency and tunes it with respect to TM_{10} mode, as observed from Fig. 2.2a. An optimum BW is obtained for 'l_s' = 6.8, 'w_s' = 0.8, and 'y' = 3.0 cm as shown in Fig. 2.2b. The simulated and measured BW is 208 MHz (23.27%) and 215 MHz (23.28%), respectively. The antenna yields a gain of above 8 dBi across the entire S_{11} BW as shown in Fig. 2.2b and as specified in [19].

To design the ANN model, a crucial step is training the ANN through the data sets. For this initially, the E-shape MSA is optimized for its fundamental TM_{10} mode frequency over a frequency range of 800–5000 MHz in a step size of 200 MHz. Here, air is used as the substrate, and top and side views of suspended E-shape MSA are shown in Fig. 2.3a, b. At each frequency, equivalent E-shape MSA is optimized for four different substrate thicknesses, i.e., 0.05λ–0.08λ. The E-shape design is carried out in the following manner for each frequency. For each substrate thickness, to create the training data set for the ANN model, the detail method of data set creation is represented in the flowchart as shown in Fig. 2.3d.

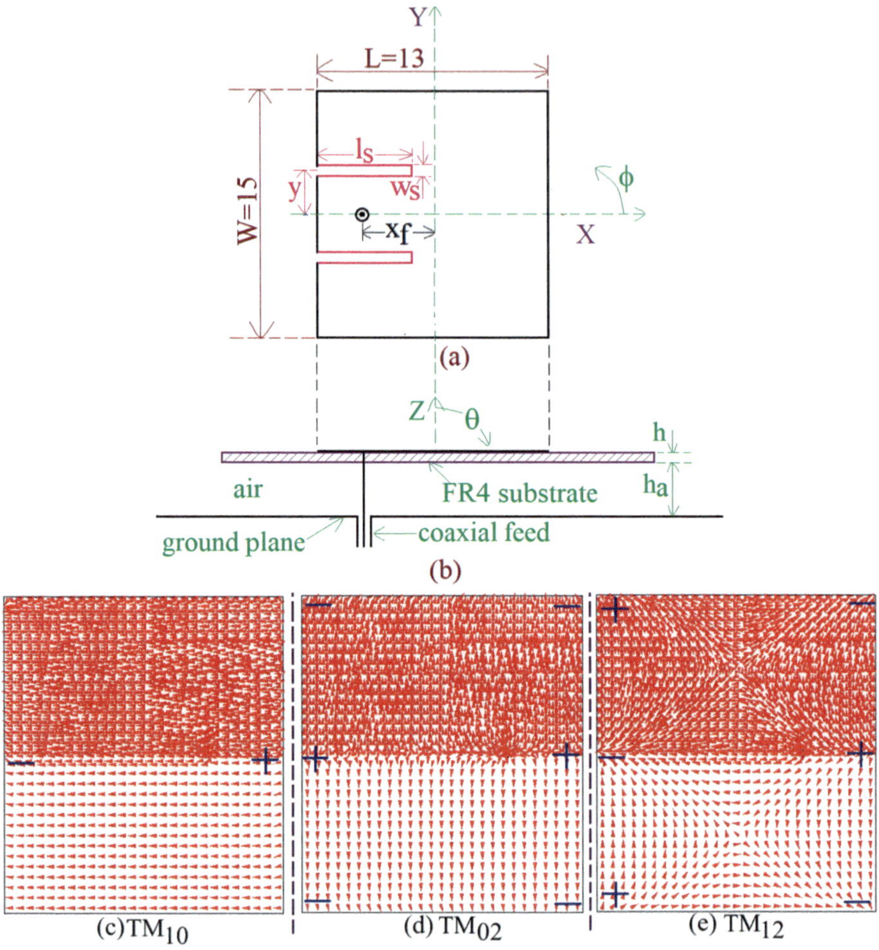

Fig. 2.1 **a**, **b** E-shape MSA [19], **c**–**e** simulated surface current distributions of various modes of RMSA for the peaks in resonance frequency plot

1. Design RMSA for a particular TM_{10} mode frequency f_r, and substrate thickness h using resonance frequency equation of RMSA [3].
2. Feed the RMSA at appropriate location starting at (L/4.0).
3. Optimize the dimensions of the RMSA (L, W) to tune the resonant peak to f_r.
4. For E-shape patch, optimize the slot dimensions l_s, w_s so that the loop due to the coupling of TM_{10} and TM_{02} modes lies inside VSWR = 2 circle.

As mentioned above, for neural network training, simulations were carried out using IE3D simulations [30] over 800–5000 MHz frequency range, in the intervals of 200 MHz and for four substrate thicknesses in the range mentioned above. With

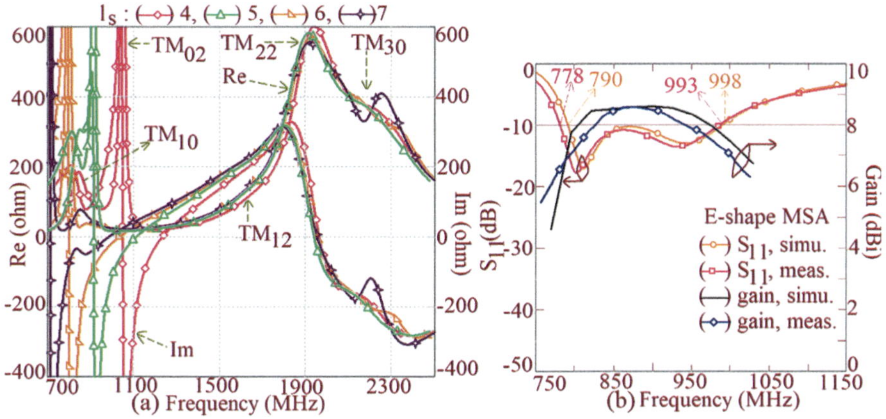

Fig. 2.2 **a** Resonance frequency plots for variations in 'l_s', **b** S_{11}, and gain plots for E-shape MSA [19]

this, in total 84 points of data to train the neural network model were generated. Sample training data set created using this procedure is given in Table 2.1.

The block diagram and detailed structure of the ANN model for the trained data are shown in Fig. 2.4a, b. It is a deep neural network model with two hidden layers and ten neurons in each hidden layer. The optimum number of layers and number of neurons can only be decided by iterative testing of the result of the ANN. The number of layers and number of neurons in each layer for an iteration which has the least error and highest accuracy must be preferred. The architecture shown in Fig. 2.4a, b is designed as two parts. These are two ANNs communicating with each other to provide accurate predictions. The details of the ANN architecture such as number of inputs, outputs, and hidden layers are specified in Table 2.2.

Initially, ANN 1 is fed with the resonant frequency ($f_r = \mathrm{TM}_{10}$), and substrate thickness (h). ANN 1 is trained to predict '$\Delta L/h$' which is the fringing length expressed in terms of substrate thickness as given by Eq. (2.1):

$$\Delta\frac{L}{h} = \left(\frac{\lambda}{2} - L\right)x\frac{1}{2h} \tag{2.1}$$

This quantity predicts the patch length (L) of the antenna. The patch width in turn can be derived using the ratio W/L observed from the optimum configuration. By predicting one value, performance of the ANN is increased. ANN 2 of the structure predicts 4 dimensions; slot length factor, slot width factor, distance from the centre of the patch to the centre of the slot width as a ratio with patch width (y/W), and x-coordinate of the feed position, 'x_f'. The slot length factor is the ratio of the length of slot to the patch length (l_s/L) and, similarly, the slot width factor is the ratio of the slot width to the patch width (w_s/W). These ratios are used to keep the range of the output in the [0, 1] range. As the sigmoid activation function is used, the output of every neuron will be in the range [0, 1]. Hence, it is better and efficient to use

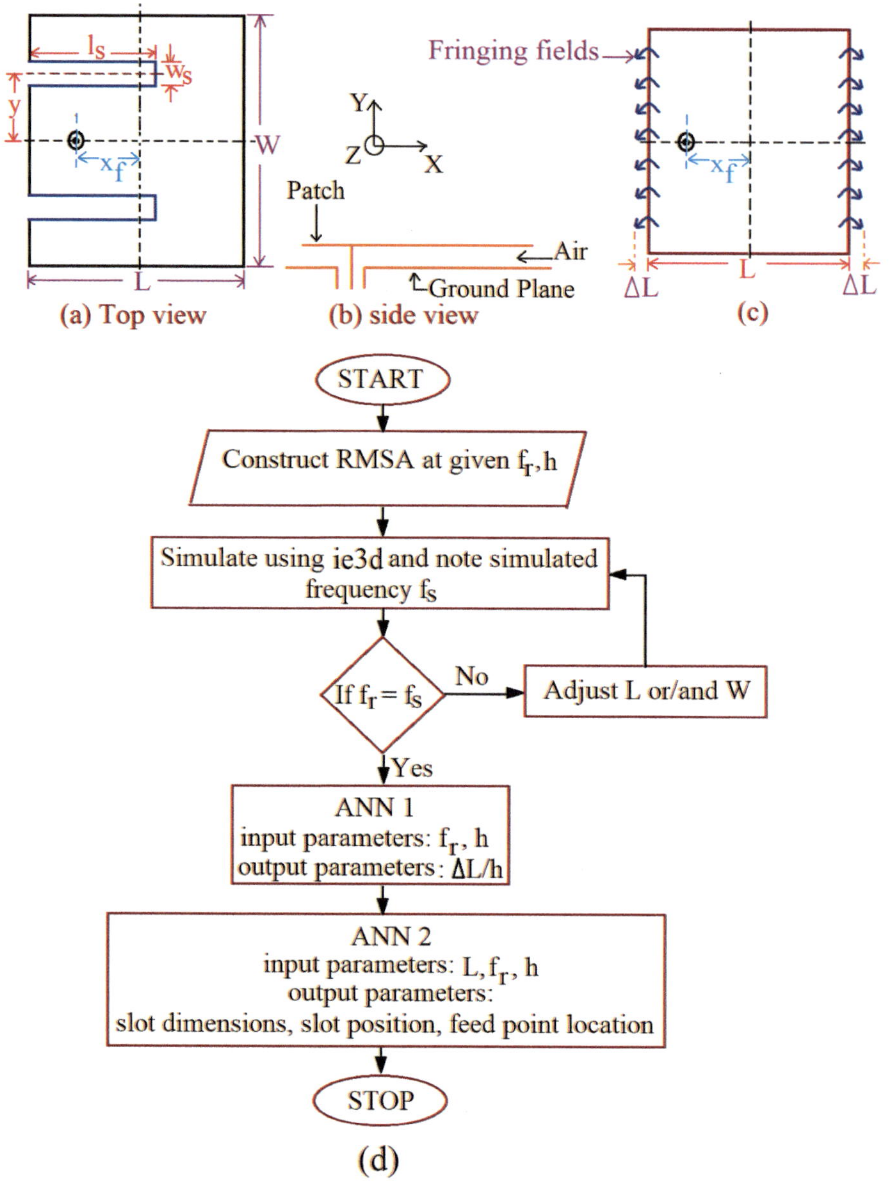

Fig. 2.3 Air-suspended E-shape MSA **a** top view, **b** side view, **c** fringing fields of RMSA, and **d** flowchart showing the design procedure of E-shape MSA for data set generation for ANN training

Table 2.1 ANN training data set for E-shape MSA

f_r	h	L	W	l_s	w_s	y	x_f
794	1.9	15.6	18.8	18.0	1	5.2	3.9
1000	2.4	11.5	13.8	18.8	1	1.8	2.9
1103	1.4	11.2	13.5	11.8	0.3	3.9	2.8
1400	1.3	8.6	10.4	10.0	0.8	2.5	2.1
1709	1.0	7.1	8.5	7.6	0.5	2.8	1.8
1998	1.1	5.8	7.0	9.0	0.5	1.3	1.4
2405	6.0	5.2	6.2	5.0	0.5	2.1	1.3
2776	7.0	4.3	5.2	5.0	0.3	1.2	1.1
3026	5.0	4.1	4.9	4.0	0.3	1.5	1.0
3384	5.0	3.6	4.3	3.8	0.3	1.8	0.9
3859	6.0	3.0	3.6	4.6	0.2	0.6	0.7
4190	6.0	2.7	3.3	4.0	0.2	0.6	0.7
4470	5.0	2.6	3.1	4.6	0.2	0.4	0.6
4759	5.0	2.4	2.9	3.4	0.2	0.6	0.6
4979	6.0	2.2	2.7	4.3	0.2	0.3	0.5

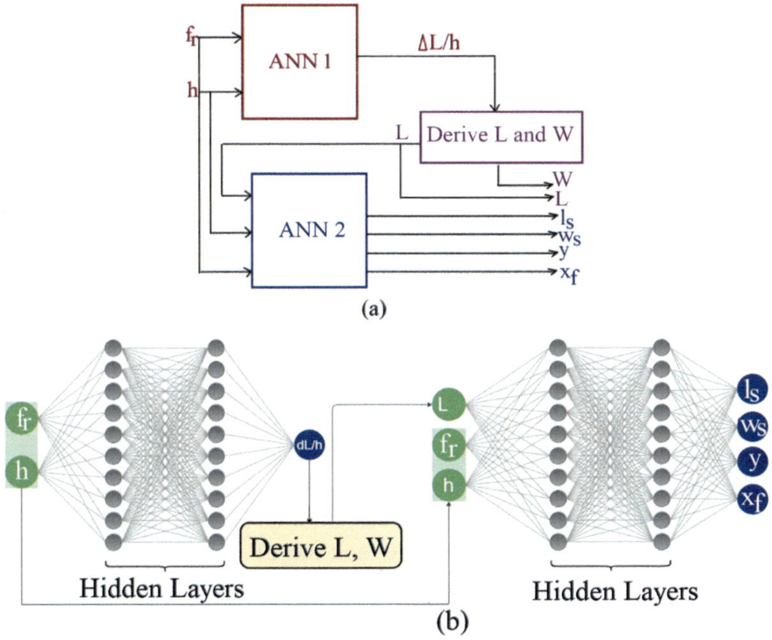

Fig. 2.4 **a** Block diagram of ANN model for trained data, **b** ANN architecture with different layers and nodes for air-suspended E-shape MSA

Table 2.2 Number of inputs and outputs in ANN

Number of inputs	2
Number of outputs	4
Number of hidden nodes	20
Number of hidden layers	2
Number of hidden nodes per layer	20/2 = 10

inputs and outputs that adhere to the range. This way the firing or activation of the neuron is true to the data set. The learning is more efficient. Further, the trained ANN structure is used to predict the dimensions of E-shape MSA for a given TM_{10} mode frequency (f_r) and substrate thickness (*h*). The dimensions predicted by the two-step ANN architecture are given in Table 2.3, after considering minor changes. These minor changes in the dimensions are done to obtain practically realizable values for each predicted E-shape parameter.

The errors of trained and predicted dimensions by ANN 1 and ANN 2 with varying epochs are shown in Fig. 2.5a, b. A reduction in error is observed with a large number of epochs. From the dimensions observed from the table, the E-shape MSA is designed, and simulations are conducted to verify the result in terms of resonance frequency peaks and VSWR plot for the BW. The results in terms of impedance BW are shown in Fig. 2.5c, d. Simulated BWs of 200 MHz (15.38%) and 380 MHz (20.99%) are realized by E-shape MSAs designed on air substrate at 1300 MHz and 1750 MHz, at $h = 0.05\lambda$ and 0.07λ respectively. The fabricated prototypes of both E-shape MSAs using air substrate are shown in Fig. 2.6a, b.

2.2.2 *ANN Model for Offset E-Shape Microstrip Antenna for Broadband and CP Response*

Over the BW, a broadside radiation pattern with a reduced cross-polar level is observed in the wideband E-shape MSA. It provides an AR value of greater than 15 dB across the BW, indicating LP. However, antennas with a greater cross-polar level or CP response are preferable in wireless applications, where signal loss from the interference and multi-path propagation effects is needed to be minimized.

The lengths of pair of slots in the wideband E-shape MSA design are equal. Therefore, an equal length offset slot cut design of an E-shape MSA is investigated [20] to obtain both the wideband response and the CP response. This is so that the offset slots can help to excite bidirectional currents on the patch to produce the CP characteristics by causing an imbalance in the structure. The configuration of offset E-shape MSA and the Smith chart showing broadband and CP response reported in [20] are shown in Fig. 2.7a, b.

To ensure broadband response as that of E-shape MSA, it was specified in [20] that one of the slots at position 'y_1' is kept as that of E-shape MSA and above coaxial feed axis, whereas the other slot position '*y*' is shifted to an offset position,

Table 2.3 Optimized and practically realizable ANN values

f_r	h	L	W	x_f	y	l_s	w_s
600	2.5	20.9	25.1	5.1	7.4	13.3	1.3
600	3.5	19.9	23.9	4.9	4.5	9.4	1.7
700	2.1	17.9	21.5	4.3	6.8	11.4	1.1
700	3.0	16.9	20.3	4.1	3.8	5.95	1.5
1100	1.4	11.2	13.4	2.7	4.1	7.25	0.7
1100	1.9	10.7	12.8	2.6	2.4	3.25	0.9
1900	0.8	6.5	7.8	1.5	2.6	4.15	0.4
1900	1.1	6.1	7.4	1.5	1.4	2.7	0.5
2300	0.7	5.3	6.3	1.2	2.0	3.35	0.3
2300	0.9	5.1	6.1	1.2	1.2	2.3	0.4
2700	0.6	4.5	5.4	1.1	1.7	3.05	0.3
2700	0.8	4.3	5.1	1.0	0.9	2	0.3
3100	0.5	3.9	4.7	0.9	1.5	2.65	0.2
3100	0.7	3.7	4.5	0.9	0.8	1.75	0.2
3500	0.4	3.5	4.3	0.8	1.4	2.3	0.2
3500	0.6	3.3	4.0	0.8	0.7	1.65	0.2
3900	0.4	3.1	3.7	0.7	1.1	1.9	0.2
3900	0.5	3.0	3.6	0.7	0.8	1.45	0.2
4300	0.3	2.9	3.5	0.7	1.2	1.95	0.2
4300	0.5	2.7	3.2	0.6	0.6	1.3	0.2
4700	0.3	2.6	3.1	0.6	1.1	1.5	0.1
4700	0.4	2.5	3.0	0.6	0.8	1.5	0.1
5100	0.3	2.5	3.0	0.6	0.8	1.4	0.1
5100	0.4	2.2	2.7	0.5	0.7	13.3	0.1

below. Against variations in 'y_1' and 'y' as mentioned in Fig. 2.7a–c, the loop in the Smith chart forms in the higher inductance area, and thus impedance locus becomes inductive. With variation in 'y_1' and 'y', effective patch width as seen by the coaxial feed increases, which is linked to an increase in the inductive component in the input impedance. An additional loop smaller in size is seen in the impedance locus with a larger spacing between the two slots. Here wideband response in the Smith chart is represented by a bigger loop [20]. The offset position of the slots in E-shape MSA tunes the spacing in between TM_{10}, TM_{02}, and TM_{11} resonant modes that achieve the wideband response followed by a CP response at a smaller loop size position [20].

Reported design given in [20] was considered on suspended FR4 substrate against that configuration selected in the present study for the application of ANN is selected on air substrate as shown in Fig. 2.8a, b. Like the E-shape MSA in the previous section, an offset E-shape MSA is optimized for its fundamental TM_{10} mode frequency over

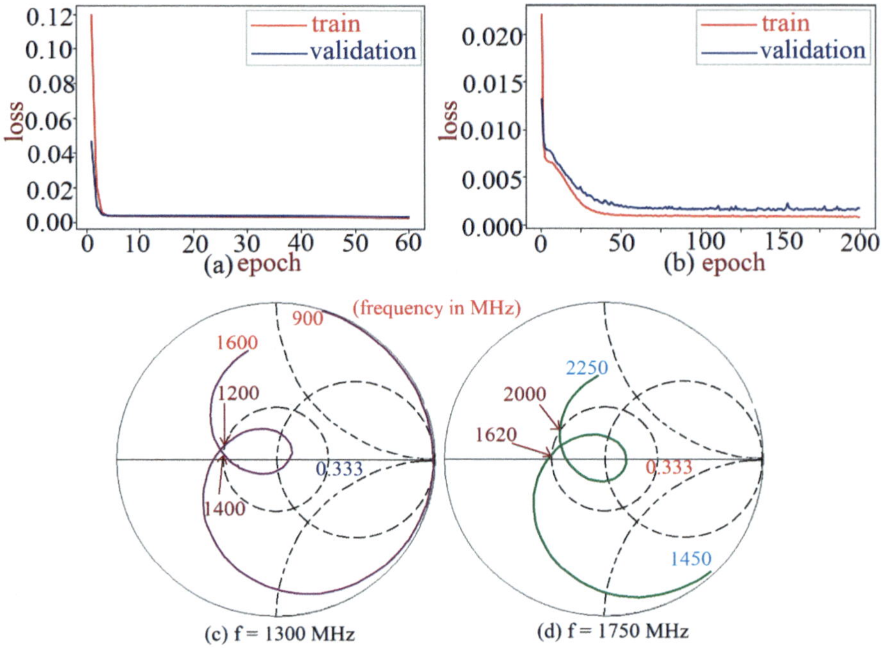

Fig. 2.5 Error versus epochs for **a** Step 1, **b** Step 2, and Smith charts of E-shape MSA at **c** h = 12 mm (0.05λ), and **d** h = 12 mm (0.07λ)

Fig. 2.6 Fabricated E-shape MSA for **a** h = 12 mm (0.05λ), **b** h = 12 mm (0.07λ)

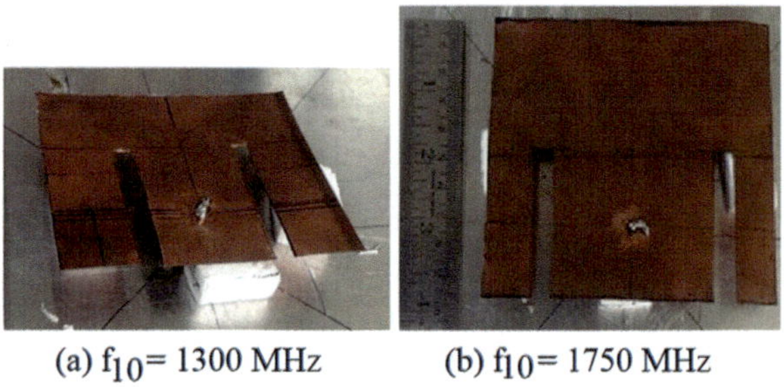

a frequency range of 800–5500 MHz. In this configuration for generating data sets, step size in frequency of 100 MHz is considered for 800–2000 MHz range and step size of 200 MHz for 2000–5500 MHz range. For each frequency, antenna designs were simulated at four different substrate thicknesses in the range of 0.05λ–0.08λ.

The initial design of RMSA for the offset E-shape MSA design is carried out in the following manner for each frequency and for each substrate thickness to create training data set for the ANN, and the detailed method of data set creation is represented in the flowchart shown in Fig. 2.3d.

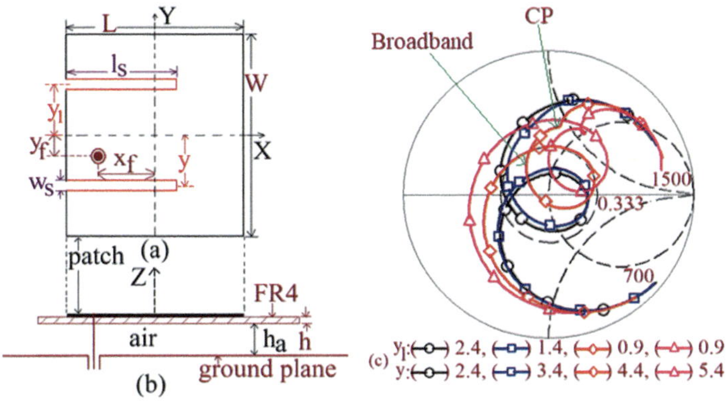

Fig. 2.7 **a**, **b** offset E-shape MSA, its **c** smith chart for varying slot positions [20]

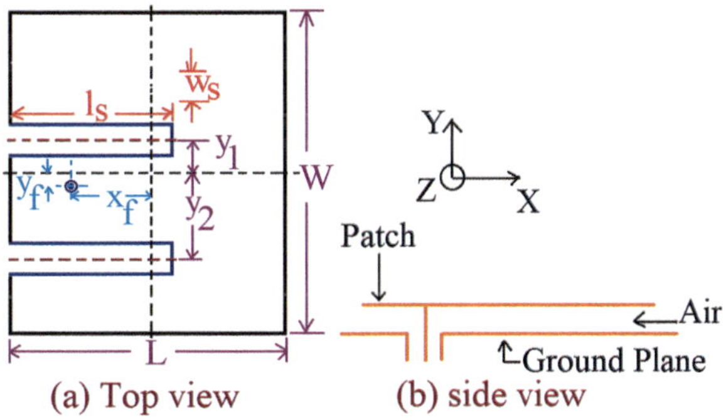

Fig. 2.8 Air-suspended offset E-shape MSA **a** top view, **b** side view

1. Design RMSA for a specific TM_{10} mode frequency f_r, and substrate thickness h using resonant frequency equation of RMSA [3].
2. Feed the RMSA at appropriate location starting at (L/4.0).
3. Optimize the dimensions (L, W) of the RMSA to tune the resonant peak to f_r.
4. Optimize the slot dimensions l_s, w_s, slot positions y_1 and y_2, and feed point location (x_f, y_f) so that the loops representing broadband and CP responses due to the coupling of TM_{10}, TM_{02}, and TM_{11} modes lie inside the VSWR = 2 circle.

The ANN model of the offset E-shape MSA is shown in Fig. 2.9. In this model ANN 1 model predicts the dimensions L and W of the RMSA like that of ANN 1 of the E-shape MSA. Part 2 of the architecture predicts 5 dimensions, slot length factor 'l_s', slot width factor 'w_s', distance from the centre of the patch to the centre of the first slot 'y_1', distance from the centre of the patch to the centre of the second slot 'y_2', and finally the x-axis coordinate of the feed position, x_f. For all designs, y-axis coordinate of the feed point is considered as 'y_f' = 0.6. The slot width factor is the ratio of the width of slot to the patch width 'w_s/W' and, similarly, the slot length factor is the ratio of the slot length to the patch length 'l_s/L'. These ratios are used to keep the range of the output in the [0, 1] range. The details of the ANN architecture

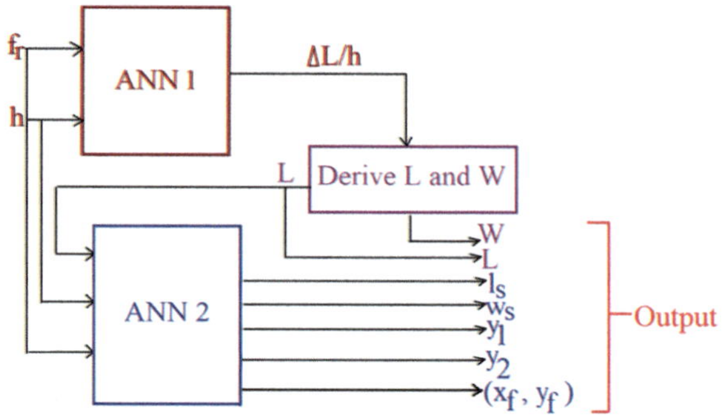

Fig. 2.9 Air-suspended offset E-shape ANN model

Table 2.4 Number of inputs and outputs in ANN architecture for offset E-shape MSA design

Number of inputs	2
Number of outputs	5
Number of hidden nodes	20
Number of hidden layers	2
Number of hidden nodes per layer	20/2 = 10

such as number of inputs, outputs, and hidden layers are specified in Table 2.4, and the structure is similar to the one shown in Fig. 2.4b.

For the training purpose of the neural network, simulations were carried out using IE3D Software over a frequency range of 800–5500 MHz. A step size of 100 MHz is selected for frequencies in the range 800–2000 MHz and 200 MHz step size for frequencies in the range 2000–5500 MHz. For each frequency, antenna designs were simulated at four different substrate heights, i.e., 0.05λ–0.08λ. In total there were 84 points of data to train the neural network model. Sample training data set created using the above procedure is given in Table 2.5.

Further, the trained ANN structure is used to predict the dimensions of offset E-shape MSA for a given TM_{10} mode frequency (f_r) and substrate thickness (h). The sample data of dimensions predicted by the two-step ANN architecture is given in Table 2.6. Minute modifications are done in these predicted dimensions to obtain practically realizable values. It can be observed that the ratios of $\Delta L/h$, l_s/L, w_s/W, y_1/W, y_2/W, and x_f/L of the predicted dimensions remain almost constant proving the consistency in the prediction by the proposed ANN model.

To verify, offset E-shape MSA designed from the ANN predicted dimensions was fabricated and experimentally verified. The various results for the antenna design at 950 MHz are shown in Fig. 2.10. The simulated and measured Smith charts, broadside gain, and AR BW plots are shown. In Smith chart, two loops representing the wideband and CP responses are noted. A simulated reflection coefficient BW of 327 MHz (31.7%), and a measured BW of 361 MHz (34.89%) are noted. Due to the excitation of orthogonal resonant modes, CP response with simulated AR BW of 30 MHz (2.6%) and measured AR BW of 36 MHz (3.1%) are observed. The

Table 2.5 ANN training data set for offset E-shape MSA

f_r	h	L	W	l_s	w_s	y_1	y_2	x_f
800	2.4	15.6	17.95	9.35	0.95	1.1	6.35	5.05
1000	1.95	12.5	14.35	7.5	0.75	0.86	5.05	4.05
1200	1.6	10.3	11.85	6.2	0.65	0.7	4.2	3.306
1400	1.392	8.8	10.12	5.28	0.5363	0.6072	3.5723	2.843
1600	1.21875	7.8	8.97	4.68	0.475	0.5382	3.166	2.52
1800	1.083	6.9	7.935	4.14	0.42	0.4761	2.801	2.229
2000	0.975	6.1	7.015	3.66	0.3717	0.4209	2.4762	1.971
2200	0.886	5.6	6.44	3.36	0.341	0.3864	2.273	1.809
2400	0.8125	5.2	5.98	3.12	0.3169	0.3588	2.1109	1.68
2800	0.696	4.5	5.175	2.7	0.274	0.3105	1.8267	1.4539
3200	0.609	3.8	4.37	2.28	0.2316	0.2622	1.5426	1.2277
3600	0.5416	3.5	4.025	2.1	0.213	0.2415	1.4208	1.1308
4000	0.48	3.1	3.565	1.86	0.1889	0.2139	1.258	1.0016
4400	0.443	2.9	3.335	1.74	0.1767	0.2001	1.177	0.9369
4800	0.40625	2.6	2.99	1.56	0.158	0.1794	1.055	0.84

peak antenna gain of about 9.2 dBi is observed in the simulation and measurement. The fabricated prototype of offset E-shape MSA designed at 950 MHz is shown in Fig. 2.10c.

2.2.3 *ANN Model for Modified E-Shape MSA with Asymmetrical Slots for CP Response*

One more configuration of modified E-shape MSA which realizes CP response is RMSA with asymmetrical slots. Here asymmetry is incorporated in the patch structure by using two slots of different lengths. Due to this inequality, resonant modes with orthogonal current distributions are excited resulting in a CP response. The geometry of E-shape MSA with asymmetric slots is shown in Fig. 2.11a, b. Initially a basic RMSA is designed for its fundamental TM_{10} mode frequency of 900 MHz. The MSA is optimized on a suspended FR4 substrate with air thickness $h_a = 1.9$ cm, making total thickness of the antenna as 2.06 cm. Antenna is fed using a 50Ω coaxial N-type connector. The RMSA dimensions are determined as $L = 12.2$ and $W = 14$ cm. It is well known that incorporation of a pair of rectangular slots of equal length results in a symmetrical E-shape MSA realizing wideband response. But, when one of the slot lengths of E-shape MSA is modified, it becomes asymmetric with respect to the proximity feed location. With respect to the surface currents on both the sides of the feed point due to the varied slot lengths, the feed point is no

Table 2.6 Predicted data set of offset E-slot RMSA for ANN

f_r	h	L	W	l_s	w_s	y_1	y_2	(x_f, y_f) (cm)
700	2.8	17.606	20.247	10.560	1.0659	1.2105	7.13092	5.66752
750	2.4	16.724	19.232	10.028	1.0138	1.1504	6.77636	5.38217
850	2.3	14.508	16.684	8.699	0.8795	0.9980	5.87814	4.66857
950	2.05	12.992	14.940	7.7894	0.7882	0.8941	5.26507	4.17981
1050	1.85	11.762	13.526	7.0512	0.7141	0.8098	4.76749	3.78323
1150	1.7	10.725	12.334	6.4293	0.6516	0.7387	4.34793	3.44913
1250	1.55	9.8881	11.371	5.9269	0.6011	0.6813	4.00925	3.17913
1350	1.45	9.1370	10.507	5.4765	0.5557	0.6297	3.7051	2.93718
1450	1.35	8.5088	9.7852	5.0998	0.5178	0.5866	3.45089	2.73479
1550	1.25	7.98	9.1770	4.7826	0.4860	0.5503	3.23707	2.56436
1750	1.1	7.083	8.1455	4.2449	0.4318	0.4886	2.87419	2.27553
1950	1.0	6.344	7.2961	3.8023	0.3870	0.4377	2.575	2.03793
2300	0.85	5.3863	6.1942	3.2282	0.3290	0.3717	2.18712	1.72982
2700	0.7	4.6334	5.3284	2.7770	0.2835	0.3197	1.88288	1.48781
3100	0.65	3.9898	4.5883	2.3917	0.2440	0.2752	1.62114	1.28127
3500	0.55	3.5774	4.1140	2.1444	0.2191	0.2466	1.45466	1.1488
3900	0.5	3.2080	3.6892	1.9231	0.1965	0.2210	1.3047	1.0303
4300	0.45	2.9188	3.3566	1.7498	0.1788	0.2010	1.1874	0.93749
4700	0.4	2.6888	3.0921	1.6118	0.1648	0.1851	1.09426	0.86367
5100	0.4	2.4396	2.8056	1.4627	0.1493	0.1678	0.99236	0.78386
5300	0.35	2.3932	2.7522	1.4346	0.1467	0.1646	0.97422	0.76887
5500	0.35	2.2906	2.6342	1.3732	0.1403	0.1575	0.93224	0.73601

more at the centre along the y-axis. This asymmetry excites orthogonal TM_{11} mode [21]. The surface currents of all the modes mentioned above are shown in Fig. 2.11c.

The length of the slots is parametrically optimized one after the other keeping the other slot length constant to tune the impedance of TM_{11} mode and bring it inside VSWR = 2 circle as shown in Fig. 2.11d. On parametric optimization, the dimensions of asymmetric E-shape MSA are identified as $l_1 = 4$, $l_2 = 6$, $w = 0.8$, $y = 2.4$, and $x_f = 2.8$ cm. However, for these slot lengths, impedance of TM_{10} and TM_{02} modes is high due to which the loop indicating coupling between these two modes lies outside VSWR = 2 circle. With slot length increment, shifting of these loops outside VSWR = 2 circle occurs. At the same time due to the created asymmetry, TM_{11} mode gets excited and the kink representing CP response gets optimized inside VSWR = 2 circle as shown in Fig. 2.11d. The optimum results representing resonant frequency plots, S_{11}, gain, and AR plots are shown in Fig. 2.11e. A simulated impedance BW of 91 MHz (8%) and AR BW of 21 MHz (1.8%) with a peak gain of 7 dBi are realized.

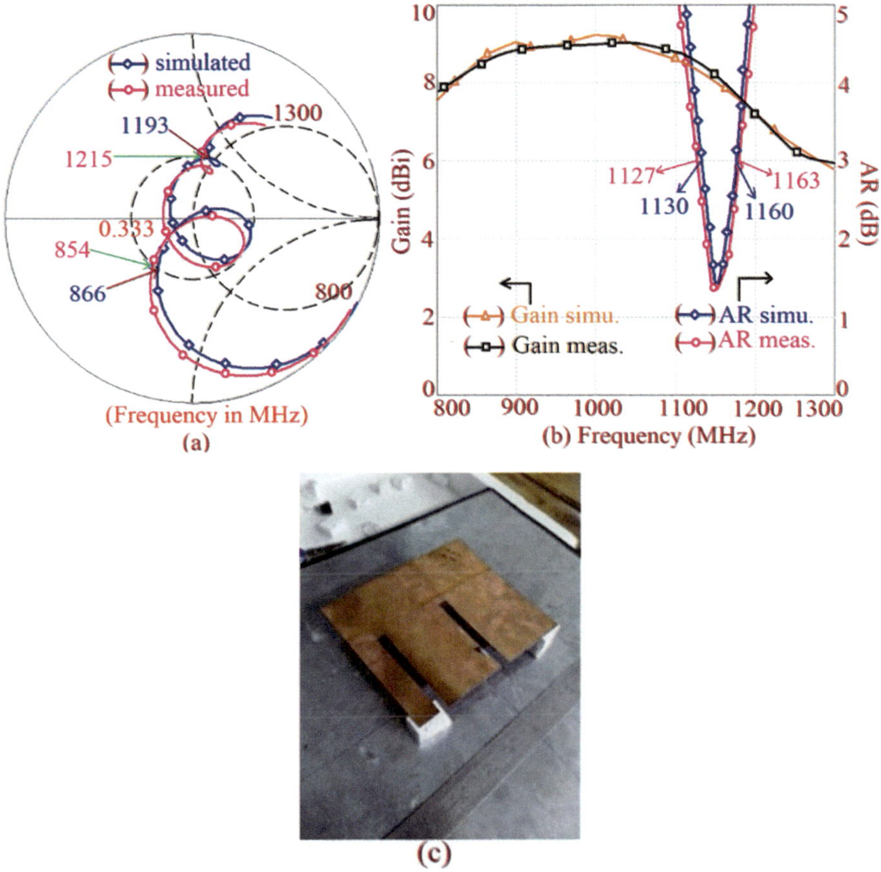

Fig. 2.10 a Smith charts representing broadband and CP responses, **b** gain and AR plots, and **c** fabricated prototype of air-suspended offset E-shape MSA designed at 950 MHz from the ANN predicted dimensions

This arrangement accomplishes a narrow impedance BW, but it realizes CP using a basic unequal length patch.

Further, an ANN model for the asymmetric E-shape MSA for CP response is developed. The architecture of the ANN model is shown in Fig. 2.12. In the modelling section, four layers are employed. The input layer is the first layer, and it consists of a single neuron with frequency characteristics. To provide nonlinearity to the model, there are two hidden layers, each with 6 neurons and a ReLU activation function, and output layer is with all 8 neurons as outputs and a sigmoid activation function. Here MinMaxScalar is adapted to scale the features to avoid a slow learning gradient descent curve for the proposed training model. As the data set has no outliers, mean squared error criterion is used for the loss function. Adam optimizer is employed in this ANN, which has both stochastic gradient descent with momentum and adaptive

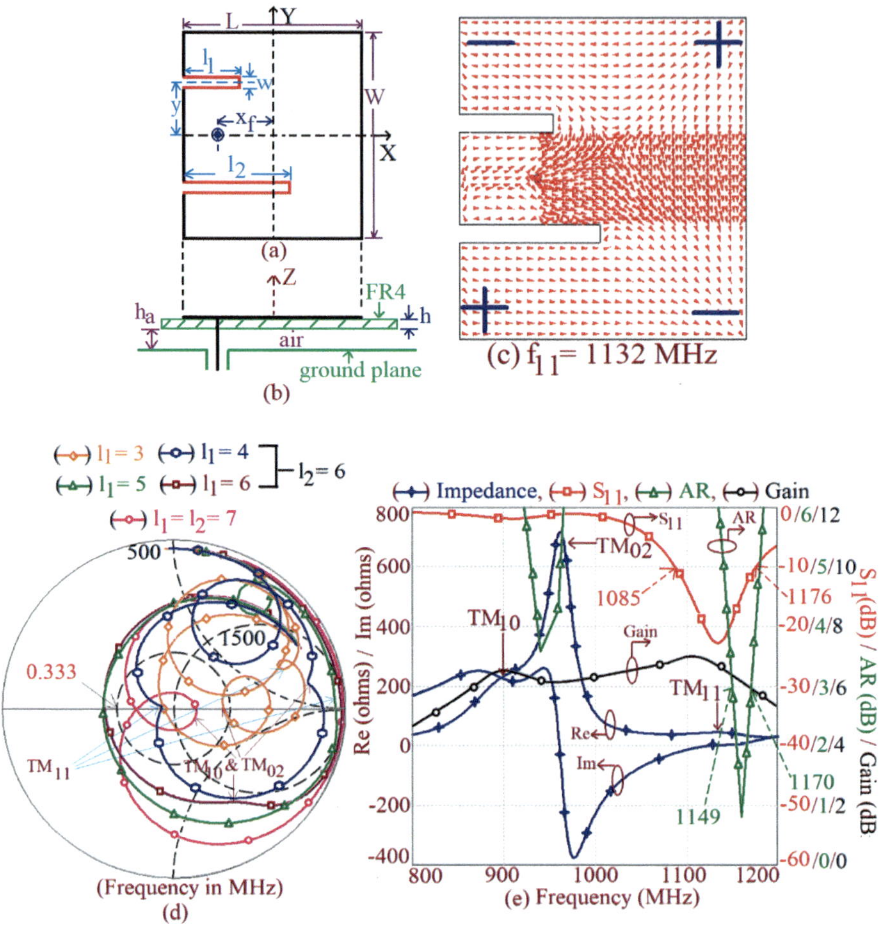

Fig. 2.11 **a**, **b** offset E-shape MSA, its **c** surface current distribution for TM_{11} mode, **d** smith chart for varying slot lengths, and **e** results of optimized configuration of asymmetric E-shape MSA for CP

learning of learning rate. The details of the ANN architecture such as number of inputs, outputs, and hidden layers are specified in Table 2.7.

The initial design of RMSA for the asymmetric E-shape MSA design is carried out in the following manner for each frequency and for the chosen substrate thickness to create training data set for the ANN, and the detailed method of data set creation is represented in the flowchart shown in Fig. 2.3d.

1. Design RMSA for a particular TM_{10} mode frequency f_r, and substrate thickness h using resonant frequency equation of RMSA [3].
2. Feed the RMSA at appropriate location starting at (L/4.0).
3. Optimize the dimensions (L, W) of the RMSA to tune the resonant peak to f_r.

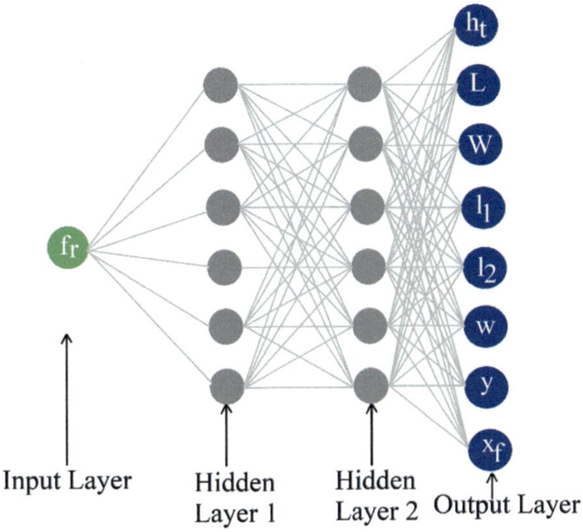

Fig. 2.12 ANN architecture with different layers and nodes for asymmetric E-shape MSA for CP response

Table 2.7 Number of inputs and outputs in ANN architecture for asymmetric E-shape MSA design

Parameter	Value
Number of inputs	1
Number of outputs	8
Number of hidden layers	2
Number of hidden neurons in both layers	6
Activation functions	ReLU for hidden layers and sigmoid for output layer
Training algorithm	ANNs

4. Further, optimize the slot dimensions l_1, l_2, w, slot positions y, and feed point location x_f so that the loops representing the CP responses due to the excitation of TM_{11} mode lie inside the VSWR=2 circle.

To create the training data set of ANN, the asymmetric E-shape MSA is optimized for its fundamental TM_{10} mode frequency over a frequency range of 700–2900 MHz in a step size of 200 MHz. The substrate height was chosen as 0.063λ for each frequency as per the initial optimum configuration designed at 900 MHz. The training data set created using the above procedure is given in Table 2.8. Further, the trained ANN structure is used to predict the dimensions of asymmetric E-shape MSA for a given TM_{10} mode frequency (f_r) and substrate thickness (h_t). The sample data of dimensions predicted by the ANN architecture is given in Table 2.9. Minute modifications are done in these predicted dimensions to obtain practically realizable values. It can be observed that the ratios of l_1/L, l_2/L, w/W, y/W, and x_f/L of the predicted dimensions remain almost constant proving the consistency in the prediction by the proposed ANN model.

Table 2.8 ANN training data set for Asymmetric E-shape MSA

f_r	h_t	L	W	$l1$	$l2$	w	y	x_f
700	2.66	15.8	18.2	5.2	7.6	1	3	3.6
900	2.06	12.2	14	4	6	0.8	2.4	2.8
1100	1.66	10	11.4	3.2	4.8	0.6	1.9	2.3
1300	1.46	8.4	9.6	2.8	4	0.5	1.6	1.9
1500	1.26	7.2	8.2	2.4	3.5	0.4	1.4	1.6
1700	1.06	6.3	7.2	2.1	3	0.4	1.2	1.4
1900	0.96	5.6	6.4	1.8	2.7	0.3	1.1	1.1
2100	0.86	5	5.8	1.7	2.4	0.3	1	1.2
2300	0.76	4.5	5.2	1.5	2.2	0.3	0.9	1.1
2500	0.7	4.2	4.8	1.4	2	0.2	0.8	1
2700	0.66	3.8	4.4	1.3	1.8	0.2	0.7	0.9
2900	0.6	3.5	4	1.2	1.7	0.2	0.7	0.8

Table 2.9 ANN predicted data set for Asymmetric E-shape MSA

f_r	h_t	L	W	l_1	l_2	w	y	x_f
800	2.46	14.2	16.4	4.7	6.9	0.9	2.7	3.2
1000	2.06	12.2	14.0	4.0	5.8	0.8	2.4	2.7
1200	1.4.6	9.4	10.6	2.9	4.2	0.6	1.8	2.2
1400	1.06	6.8	7.6	2.0	3.0	0.4	1.2	1.6
1600	0.96	6.0	6.8	1.9	2.8	0.4	1.1	1.0
1800	0.9	5.4	6.2	1.8	2.7	0.3	1.0	1.0
2000	0.86	5.0	5.8	1.7	2.6	0.3	1.0	1.0
2200	0.8	4.7	5.4	1.6	2.5	0.3	0.9	1.0
2400	0.76	4.5	5.2	1.6	2.4	0.25	0.9	1.0
2600	0.76	4.4	5.1	1.5	2.3	0.2	0.9	1.0
2800	0.7	4.3	5.0	1.5	2.3	0.2	0.8	1.0

To verify, the asymmetric E-shape MSA designed from the ANN predicted dimensions was fabricated and experimentally verified. The simulated Smith chart with a kink representing CP response is shown in Fig. 2.13a. The S_{11}, gain, and AR BW plots of the offset E-shape MSA designed at 800 MHz from the ANN predicted dimensions are shown in Fig. 2.13b. A simulated impedance BW of 77.8 MHz (8%) and a measured BW of 84 MHz (8.6%) are realized. Similarly, due to the excitation of orthogonal TM_{11} mode, CP response with simulated AR BW of 16 MHz (1.5%) and measured ARBW of 20 MHz (1.9%) with a peak gain of about 7 dBi are observed. The fabricated prototype of asymmetric E-shape MSA designed at 800 MHz is shown in Fig. 2.14.

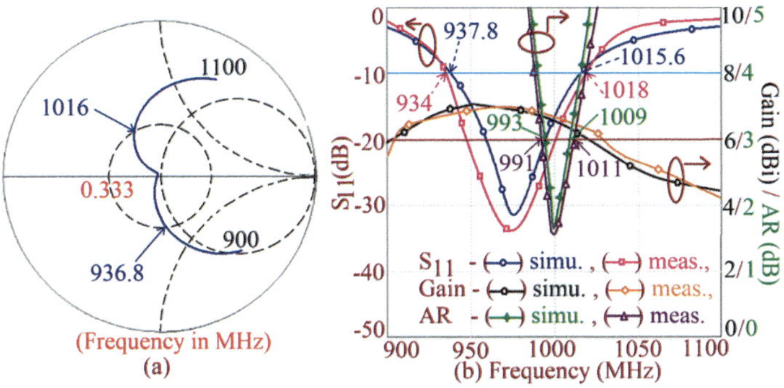

Fig. 2.13 **a** Smith chart representing CP response, simulated and measured **b** S_{11}, gain, and AR plots asymmetric E-shape MSA designed at 800 MHz on suspended substrate from the ANN predicted dimensions

Fig. 2.14 Fabricated prototype of asymmetric E-shape MSA designed at 800 MHz on suspended substrate from the ANN predicted dimensions

2.2.4 ANN Model for Multiple Slot Cut RMSA for Dual Band CP Response

Resonant slot cut MSAs are the optimum designs to yield CP response, offering reasonably higher value of the AR BW and gain. These characteristics are attributed to their single patch thicker substrate configuration. Broadside radiation pattern with a linear polarization over the impedance BW is realized in slot cut broadband MSAs by means of uni-directional modal surface currents on the patch, which are guaranteed by the symmetrical slot lengths and location. In response to this, slot settings in resonant slot cut CP MSAs guarantee surface current fluctuation along both the dimensions, which helps to generate CP wave [24–26]. Slot cut MSAs for single-band CP response were first published in the literature and here either a wideband

CP response [27] or a CP response in two bands [28] has been obtained by inserting more slots in the slot cut MSA. Higher order resonant modes in the multiple slot cut CP MSAs provide circular polarized features, and this is acquired in the direction away from the broadside [27, 28]. This creates a challenge for the antenna's alignment with the incoming radio waves. The higher order mode currents throughout the patch width and length are changed when the patch is sliced into several slots along each of its dimensions [28]. The CP response is evident away from the broadside due to these changed higher order modal currents along the two patch dimensions. Slots will however mostly change the current components in one direction and have little effect on the currents in the other direction if they are only present along one dimension in the patch. Furthermore, the CP response in the broadside direction may be obtained by investigating two-dimensional vector current variation on the slot cut patch at its modified higher order resonant modes like TM_{12} mode.

To provide dual band features, more than one resonant mode must be tuned [28]. To achieve dual band CP response in the broadside direction, multiple slot cut design of RMSA is investigated here based on the above insights about the slot selection, as shown in Fig. 2.15a, b. Three rectangular slots of varying lengths are cut on one of the patch edges in the suggested design, and an offset coaxial feed location is considered. Three slots tune more than one higher order modes on the slot cut RMSA with reference to fundamental mode, and offset feed position strengthens the bidirectional current variation across the patch to yield CP characteristics. Initially, using air-suspended low-cost FR4 substrate ($\varepsilon_r = 4.3$, $h = 0.16$ cm) and IE3D simulation [30], RMSA length 'L' is optimized for $f_{TM10} = 1100$ MHz, with an air gap of $h_a = 1.8$ cm. The simulation yields a length of 9.8 cm and a width of 12.2 cm. This results in $W/L = 1.245$, which when utilizing a single patch design realizes a modest value of the gain [3]. Parametric changes for three rectangular slot lengths are carried out to optimize the antenna for dual band CP response. The antenna yields optimum dual band response for $y_1 = 0.8$, $y_2 = 2.5$, $y_3 = 4.2$, $l_1 = 8.3$, $l_2 = 9.1$, $l_3 = 7.3$, $w_s = 0.9$, $x_f = 2.6$, and $y_f = 3.3$ cm, as shown in Fig. 2.15c. The simulated and measured results indicate that the impedance BW is 665 MHz (49.83%) and 685 MHz (51.87%), respectively. The two CP BWs in the dual bands that exhibit AR values less than 3 dB in simulation are 26 MHz (2.05%) and 73 MHz (4.59%). The measurement yielded the respective AR BW of 28 MHz (2.24%) and 69 MHz (4.39%).

The ANN model for the multiple slot cut RMSA for dual band CP response is like the one shown in Fig. 2.16. This model consists of four layers out of which two are hidden layers with six neurons in each layer and a ReLU activation function. Input layer consists of single neuron of frequency characteristics, and output layer is with all 12 neurons as outputs and a sigmoid activation function. Since the data set contains no outliers, the mean squared error criterion is used for the loss function. Adam optimizer is used in this ANN, which has both stochastic gradient descent with momentum and adaptive learning of learning rate. Here, MinMaxScalar is adapted to scale the features to avoid a slow learning gradient descent curve for our training model. The details of the ANN architecture such as number of inputs, outputs, and hidden layers are specified in Table 2.10.

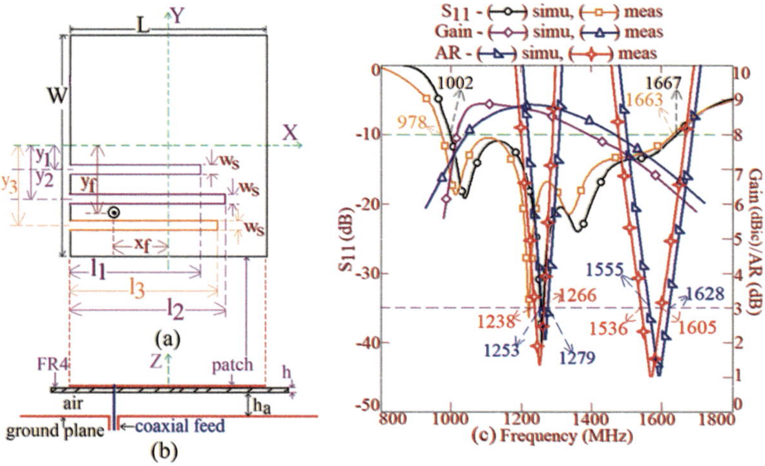

Fig. 2.15 **a, b** Multiple slot cut MSA for dual band CP response, and **c** S_{11}, AR BW, and gain plots for design at TM_{10} mode frequency of 1100 MHz

Fig. 2.16 ANN architecture with different layers and nodes for multiple slot cut RMSA for dual band CP response

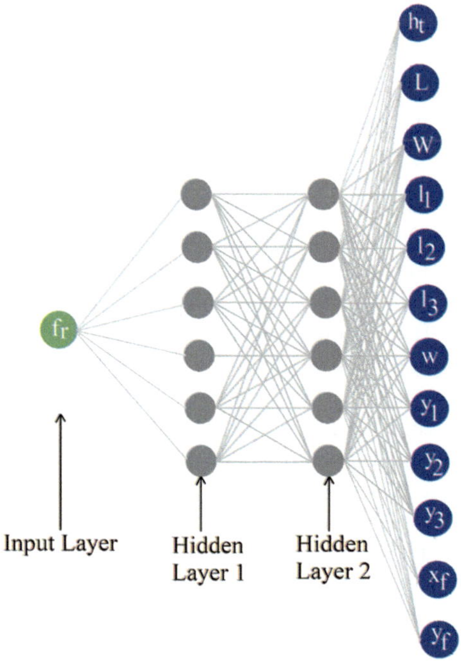

Table 2.10 Number of inputs and outputs in ANN architecture for multiple slot cut RMSA

Parameter	Value
Number of inputs	1
Number of outputs	12
Number of hidden layers	2
Number of hidden neurons in both layers	6
Activation functions	ReLU for hidden layers and sigmoid for output layer
Training algorithm	ANNs

The initial design of RMSA for the multiple slot cut RMSA design is carried out in the following manner for each frequency and for the chosen substrate thickness to create training data set for the ANN, and the detailed method of data set creation is represented in the flowchart shown in Fig. 2.3d.

1. Design RMSA for a particular TM_{10} mode frequency $f_{r,}$ and substrate thickness h using resonant frequency equation of RMSA [3].
2. Feed the RMSA at appropriate location starting at (L/4.0).
3. Optimize the dimensions (L, W) of the RMSA to tune the resonant peak to f_r.
4. Further, optimize the slot dimensions l_1, l_2, l_3, w, slot positions y_1, y_2, and y_3, and feed point location (x_f, y_f) so that the loops representing the CP responses lie inside the VSWR = 2 circle.

To create the training data set of ANN, the multiple slot cut RMSA is optimized for its fundamental TM_{10} mode over a frequency range of 700–2900 MHz in a step size of 200 MHz. The substrate height was chosen as $0.0607\lambda_g$ for each frequency as per the optimum configuration designed at 1100 MHz. The air-suspended substrate is used in this dual band design. The training data set created using the above procedure is given in Table 2.11. Further, the trained ANN structure is used to predict the dimensions of multiples slots cut RMSA for a given TM_{10} mode frequency (f_r) and substrate thickness (h_t). The sample data of dimensions predicted by the ANN architecture is given in Table 2.12. Minute variations are done in these predicted dimensions to obtain practically realizable values. It can be observed that the ratios of l_1/L, l_2/L, l_3/L, w/W, y_1/W, y_2/W, y_3/W, x_f/L, and y_f/W of the predicted dimensions remain almost constant proving the consistency in the prediction by the proposed ANN model.

To verify, multiple slot cut RMSA designed from the ANN predicted dimensions was fabricated and experimentally verified. The simulated Smith chart representing dual band CP response is shown in Fig. 2.17a. The reflection coefficient BW, gain, and AR BW plots of the multiple slot cut RMSA for dual band CP response designed at 800 MHz from the ANN predicted dimensions are shown in Fig. 2.17b. A simulated impedance BW of 535 MHz (45.98%) and a measured BW of 572 MHz (49%) are realized due to the coupling between TM_{10}, TM_{11}, and TM_{12} modes. Similarly, due to the excitation of orthogonal TM_{11} and TM_{12} modes, dual band CP response with simulated AR BW of 19 MHz (1.6%) and measured AR BW of 32 MHz (2.8%) in the first band, and simulated AR BW of 26 MHz (1.8%) and a measured value of

Table 2.11 ANN training data set for multiple slot cut RMSA

f_r	h_t	L	W	l_1	y_1	l_2	y_2	l_3	y_3	w_s	x_f	y_f
700	2.5	12.8	16	11.2	1	12	3.1	9.6	5.5	0.9	3.6	4.3
900	1.96	9.9	12.3	8.7	0.8	9.2	2.4	7.4	4.2	0.9	2.6	3.3
1100	1.6	8	10	7.1	0.7	7.4	2	6	3.5	0.7	2.1	2.7
1300	1.36	6.8	8.4	6	0.6	6.3	1.7	5	2.9	0.6	1.9	2.3
1500	1.16	5.8	7.2	5.1	0.5	5.4	1.4	4.3	2.5	0.5	1.5	2
1700	1.06	5.1	6.3	4.5	0.4	4.7	1.2	3.8	2.2	0.5	1.3	1.7
1900	0.9	4.5	5.6	4	0.4	4.2	1.1	3.4	1.9	0.4	1	1.5
2100	0.86	4.1	5	3.6	0.3	3.8	1	3	1.7	0.4	1.2	1.4
2300	0.76	3.7	4.6	3.2	0.3	3.4	0.9	2.7	1.6	0.3	1.1	1.2
2500	0.7	3.4	4.2	2.9	0.3	3.1	0.8	2.5	1.4	0.3	0.9	1.1
2700	0.66	3.1	3.8	2.7	0.3	2.9	0.8	2.3	1.3	0.3	0.8	1.05
2900	0.6	2.8	3.5	2.5	0.2	2.6	0.7	2.1	1.2	0.3	0.8	0.95

Table 2.12 ANN predicted data set for multiple slot cut RMSA

f_r	h_t	L	W	l_1	y_1	l_2	y_2	W	l_3	y_3	w_s	x_f	y_f
800	2.16	11	13.6	9.6	0.9	10.3	2.7	8.3	4.7	3.0	1.0	3.0	3.7
1000	1.96	9.8	12.1	8.6	0.8	9.1	2.4	7.3	4.2	2.6	0.9	2.6	3.3
1200	1.66	8.2	10.1	7.2	0.7	7.6	2.0	6.1	3.5	2.2	0.7	2.2	2.7
1400	1.36	6.5	8.1	5.8	0.5	6.1	1.6	4.8	2.8	1.8	0.6	1.8	2.2
1600	1.1	5.3	6.5	4.7	0.4	4.9	1.3	3.9	2.3	1.5	0.5	1.5	1.8
1800	0.9	4.4	5.4	3.9	0.3	4.1	1.1	3.2	1.9	1.3	0.4	1.3	1.5
2000	0.8	3.9	4.8	3.4	0.3	3.6	1.0	2.9	1.7	1.1	0.4	1.1	1.3
2200	0.76	3.7	4.5	3.2	0.3	3.4	0.9	2.7	1.5	1.0	0.3	1.0	1.2
2400	0.76	3.5	4.4	3.0	0.3	3.2	0.8	2.6	1.5	1.0	0.3	1.0	1.2
2600	0.7	3.5	4.3	3.0	0.3	3.2	0.8	2.5	1.4	0.9	0.3	0.9	1.1
2800	0.7	3.4	4.2	2.9	0.3	3.1	0.8	2.5	1.4	0.9	0.3	0.9	1.1
3000	0.7	3.4	4.2	2.9	0.3	3.1	0.8	2.5	1.4	0.9	0.3	0.9	1.1

44 MHz (3%) in the second band with a peak gain of about 8.43 dBi are observed. The fabricated prototype of multiple slot cut RMSA for broadband and dual band CP response designed at 800 MHz is shown in Fig. 2.17c.

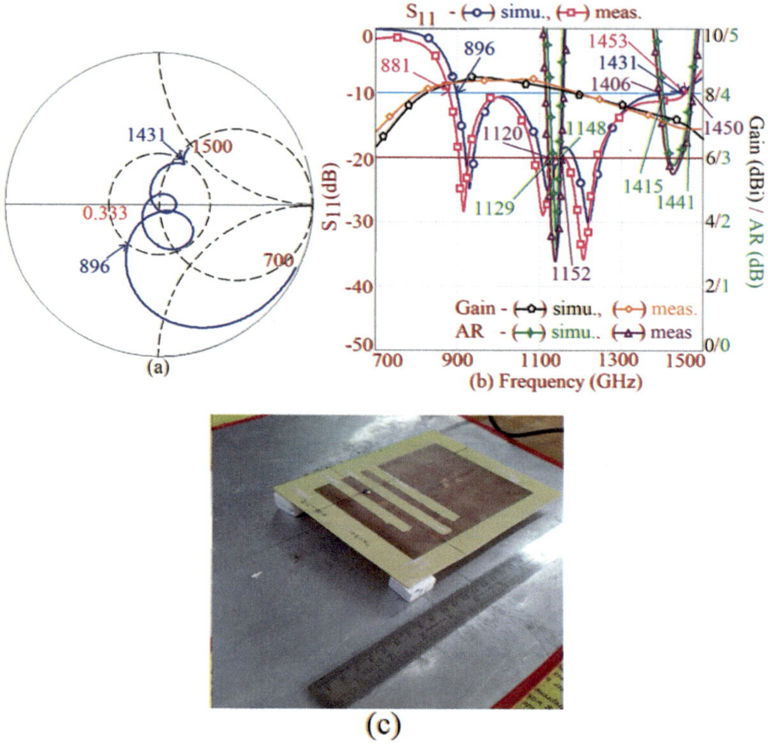

Fig. 2.17 **a** Smith chart representing CP response, simulated and measured **b** S_{11}, gain, and AR plots, and **c** fabricated prototype of multiple slot cut RMSA for broadband and dual band CP response designed at 800 MHz on suspended substrate from the ANN predicted dimensions

2.3 Summary

ANN models for air-suspended variations of rectangular slots cut MSAs such as regular E-shape MSA for broadband response, offset E-shape MSA for broadband and CP response, asymmetric slot-loaded RMSA for CP response, and multiple slot cut RMSA with offset feed for dual band CP response are presented. For all MSAs initially, its working in terms of their resonance modes is explained followed by its ANN model is developed. ANN models for all antennas used ReLU for hidden layers, and Sigmoid for the output layer. Adam optimizer is used in ANNs, which has both stochastic gradient descent with momentum and adaptive learning of learning rate, and MinMaxScalar is adapted to scale the features to avoid a slow learning gradient descent curve for our training model. In all the MSAs, frequencies obtained using ANN predicted dimensions and calculated dimensions have shown less than 5% variation. Further antenna designs using these ANN, predicted dimensions shows similar antenna response and thus proposed ANN models for wideband and CP configurations are helpful in designing similar antennas over a wide frequency range as well as substrate thickness.

References

1. Huynh T, Lee KF (1995) Single-layer single-patch wideband microstrip antenna. Electron Lett 31(16):1310–1312
2. James JR, Hall PS, Handbook of microstrip antennas, vol. 1, London Peter Peregrinus
3. Kumar G, Ray KP (2003) Broadband microstrip antennas. Artech House, USA
4. Wong KL, Hsu WH (2001) A broadband rectangular patch antenna with a pair of wide slits. IEEE Trans Antennas Propag 49(9):1345–1347
5. Fang F, Zhang XX, Ye X, Samii YR (2001) Wide-band E-shaped patch antennas for wireless communication. IEEE Trans Antennas Propag 49(7):1094–1100
6. Pandey VK, Vishvakarma VR (2007) Analysis of an E-shaped patch antenna. Microw Opt Technol Lett 49(1):4–7
7. Ge Y, Esselle KP, Bird TS (2006) A compact E-shaped patch antenna with corrugated wings. IEEE Trans Antennas Propag 54(8):2411–2413
8. Ooi BL, Shen Q (2005) A novel E-shaped broadband microstrip patch antenna. Microw Opt Technol Lett 27(5):348–352
9. Neves ES, Heckler MVT, Schildberg R, Lacava JC, Cividanes L (2003) Single layer rectangular patch antenna with two pair of parallel slits. Microw Opt Technol Lett 37(5):355–359
10. Chung KL, Wong CH (2009) Wang-shaped patch antenna for wireless communications. IEEE Antenn Wirel Propag Lett 8:468–471
11. Sharma SK, Shafai L (2009) Performance of a novel Ψ-shaped microstrip patch antenna with wide bandwidth. IEEE Antenn Wirel Propag Lett 8:468–471
12. Deshmukh A, Ray KP (2013) Analysis of broadband Ψ-shaped microstrip antennas. IEEE Antennas Propag Magaz 55(2):107–123
13. Mathew PK, Mohan S (2014) A double E-shaped microstrip patch antenna for multiband applications. ICTACT J Commun Technol 5(2):941–946
14. Babu KJ, Rama Krishna KS, Pratap Reddy L (2010) A modified E-shape patch antenna for mimo systems. Int J Comput Sci Eng 2(7):2427–2430
15. Deshmukh A, Ray KP (2011) Broadband proximity fed modified rectangular microstrip antennas. IEEE Magaz Antennas Propag 53:41–56
16. Islam MT, Shakib MN, Misran N (2009) Multi-slotted microstrip patch antenna for wireless communication. Progress Electromagn Res Lett 10:11–18
17. Deshmukh A, Kumar G (2005) Compact broadband E-shaped microstrip antennas. Electron Lett 41(18):989–990
18. Chair R, Mak CL, Lee KF, Luk KM, Kishk AA (2005) Miniaturized wide-band half U-slot and half E-shaped patch antennas. IEEE Trans Antennas Propag 53(8):2645–2652
19. Chavali VA, Deshmukh AA (2020) Modified variations of E-shape microstrip antennas for wideband response. Int J Microwave Opt Technol 15(6)
20. Chavali VA, Deshmukh AA (2022) Wideband and circularly polarized designs of modified E-shape microstrip antennas for GSM and GPS applications. Progress Electromagn Res C 121
21. Deshmukh A, Singh D, Ray KP (2019) Modified designs of broadband E-shape microstrip antennas. Sadhana—Academy Proceedings in Engineering Science, Springer Publication, pp 44–64
22. Deshmukh A, Ray KP (2015) Analysis of broadband variations of U-slot cut rectangular microstrip antennas. IEEE Antennas Propag Mag 57(2):181–193
23. Deshmukh A, Ray KP (2017) Analysis and design of broadband U-slot cut rectangular microstrip antennas. Sadhana—Academy Proceedings in Engineering Science, Springer Publication 42(10):1671–1684

24. Lam KY, Luk K-M, Lee KF, Wong H, Ng KB (2011) Small circularly polarized U-slot wideband patch antenna. IEEE Antennas Wireless Propag Lett 10:87–90
25. Khidre A, Lee KF, Yang F, Elsherbeni A (2010) Wideband circularly polarized E-shaped patch antenna for wireless applications. IEEE Antennas Propag Magaz 52(5):219–229
26. Khidre A, Lee K-F, Yang F, Elsherbeni AZ (2013) Circular polarization reconfigurable wideband E-shaped patch antenna for wireless applications. IEEE Trans Antennas Propag 61(2):960–964
27. Mondal T, Maity S, Ghatak R, Chaudhuri, Chaudhuri SRB (2018) Design and analysis of a wideband circularly polarized perturbed psi-shaped antenna. IET Microwaves, Antennas Propag 12(9):1582–1586
28. Deshmukh AA, Odhekar AA (2021) Dual band circularly polarized modified ψ-shape microstrip antenna. Progress Electromagn Res C 115:161–174
29. CST Microwave Studio, Version 2019
30. IE3D simulation software, Version 12

Chapter 3
ANN Model of U-Slot and Modified U-Slot Microstrip Antennas for Broadband and CP Response

Abstract This chapter explores the use of artificial neural network to model and design air-suspended microstrip antennas with U-slot configurations for broadband and circularly polarized responses. The slot cut approach enhances antenna bandwidth while maintaining a low-profile design, with U-slots serving as both inductance compensators and additional resonant mode generators. Despite extensive literature on U-slot and E-shape microstrip antennas, simplified methodologies for predicting key design parameters such as slot dimensions and feed point location across wide frequency ranges and substrate thicknesses are limited. This study addresses that gap by developing artificial neural network models for various U-slot-based rectangular microstrip antennas, including standard U-slot rectangular microstrip antenna for broadband response, asymmetric U-slot square microstrip antenna for dual band and circular polarization response, and corner-truncated U-slot square microstrip antenna for enhanced circular polarization performance. All designs utilize a suspended FR4 substrate over an air gap, and the artificial neural network is trained on extensive data sets spanning 600–6000 MHz and substrate thicknesses from 0.02 to $0.1\lambda_g$. Training parameters include patch dimensions, slot geometry, and feed location obtained via simulation. The Python-based artificial neural network predicts antenna dimensions with high accuracy, achieving less than 2% error in resonance frequency. This approach offers an efficient and scalable solution for the optimized design of wideband U-slot microstrip antennas.

3.1 Introduction

The slot cut approach preserves the low-profile design, while increasing the antenna BW. First published on electrically thicker substrate, U-slot and two rectangular slot cut designs of RMSA were described in [1, 2]. According to those studies, the slot serves as a probe inductance compensator for a thicker substrate in addition to adding an extra resonant mode that expands the reflection coefficient (S_{11}) BW. It was discovered that the surface currents at the mode created by the slot were revolving around the slot length, thus adding to the inductance in the equivalent

V. A. P. Chavali et al., *Neural Network Modelling of Microstrip Antennas*, Studies in Smart Technologies, https://doi.org/10.1007/978-981-96-9431-0_3

circuit. An analogous circuit model for U-slot cut RMSA [3] is created in light of this. There are reported designs of modified E-shape MSA that either change the comparable slot configuration or add corrugations along the patch width [4, 5]. The investigation on multiple resonance behaviour for U-slot cut RMSA, which was first reported in [1], is showcased in [5]. The use of U-slot in wideband and multi-band designs for RMSA and circular MSA, excited by either coaxial or L-probe feed, is documented; and the L-probe design provides a maximum BW [6–10]. Enhancement in the BW of U-slot cut RMSA can be achieved with two U-slots or offset U-slots; the two U-slot cut configuration produces the best results [11–15]. In [16], the wideband E-shape and U-slot cut RMSAs are compared with the C-shape and double C-shape MSAs. Resonant length formulation for U-slot mode based on half-wavelength approximation is described in [17, 18]. The compact and wideband designs of E-shape and U-slot cut configurations are published in [19–21] while utilizing their symmetry across the feed point axis. Description about the wideband response in U-slot and E-shape MSA designs to provide an understanding in terms of the patch's modified resonant modes and the ensuing resonant length formulations for the patch modes is discussed in [22–25]. Thus, a lot of literature is available on the basic and modified designs of U-slot cut configurations and their design methodologies. But an easy design approach that predicts the dimensions of the U-slot cut MSAs such as slot length, slot width, their aspect ratio, and the feed point location for a wide range of frequencies and substrate thicknesses is not available in the literature.

This chapter presents the ANN model for air-suspended variations of RMSA with U-slot for broadband response, SMSA with asymmetric U-slot for dual band and CP response, and corner-truncated SMSA with U-slot for dual band and CP response. In the case of suspended dielectric substrate, low-cost and lossy FR4 substrate ($\varepsilon_r = 4.3$, $h = 0.16$ cm, $\tan\delta = 0.02$) located above the ground plane using finite air gap (h_a) has been used. Initially, the functioning of all the three antennas with respect to various resonant modes leading to broadband and/or CP response is explained followed by its ANN model in detail. In the ANN modelling for each configuration, a neural network is trained using various data sets of large frequency ranges and substrate thicknesses. These data sets include various parameters of antenna such as thickness of the substrate, resonance frequency as a function of operating wavelength, total thickness of the substrate, patch dimensions such as length, width, slot length, slot position, and feed point location obtained using simulation for the given resonance frequency, and the half-wavelength value at the considered frequency. The data sets for training ANN are included at the frequency intervals of 200 MHz over 600–6000 MHz frequency band. The ANN code is written using programming language 'Python'. Next utilizing the trained ANN model, dimensions of the U-slot MSAs are predicted over substrate thicknesses in the range of 0.02–$0.1\lambda_g$ and across 600–6000 MHz frequency spectrum. All dimensions in the tables throughout the chapter are specified in cm, and frequencies in MHz.

3.2 ANN Models for U-Slot Cut and Modified U-Slot Cut RMSAs for Broadband and CP Response

3.2.1 U-Slot Cut MSA

Design of RMSA with a U-slot is a configuration that provides enhancement in the BW, as shown in Fig. 3.1a, b. The U-slot consists of two horizontal slots and a vertical slot, and detailed working of the U-slot cut MSA is reported in [26]. As explained in it, the wideband response of U-slot cut MSA is due to the optimum spacing between TM_{10} and TM_{02} resonant modes against the variation of vertical and horizontal slot lengths. With the introduction of vertical slot and its length L_v variation, TM_{10} mode reduces slightly as it is in the orthogonal direction to the flow of TM_{10} mode currents. The introduction of horizontal slots and variation in their length reduce TM_{02} mode frequency because of the orthogonal orientation of the slots with TM_{02} mode currents. The simulated resonance frequency plots for the slot length variation are shown in Fig. 3.1c, d. The dimensions of RMSA and U-slot are considered as given in [26]. As air dielectric is considered here, the total substrate thickness $h_t = h + h_a$ is considered as 'h' here. For U-slot dimension of $L_h = 7$ and $L_v = 4$ cm, and feed point location of $x_f = 2$ cm, optimum coupling between the TM_{10} and TM_{02} resonant modes is realized, due to which the impedance locus shifts inside VSWR $= 2$ circle, as shown in Fig. 3.2a.

Simulated surface current distributions at various resonant modes are shown in Fig. 3.1e, f. The uni-directional current variation across two modes provides linear polarized wideband response. A simulated BW of about 136 MHz (20.6%) with a peak gain of 8.79 dBi is realized as shown in Smith chart in Fig. 3.2a, and S_{11} and gain plots in Fig. 3.2b. Further, to design the ANN model, the crucial step is training the ANN through data sets. For this, initially the U-slot cut MSA is optimized for its fundamental TM_{10} mode over a frequency range of 800–5000 MHz in a step size of 200 MHz. In this simulation, air-suspended patch is used. At each frequency, U-slot cut MSA is optimized for substrate heights varying from 0.05λ to 0.1λ. The U-slot cut RMSA design is carried out in the following manner for each frequency and for each substrate thickness, to create the training data set for the neural network model.

1. Design RMSA for a particular TM_{10} mode frequency f_r, and substrate thickness h using resonance frequency equation of RMSA [27].
2. Feed the RMSA at an appropriate location starting at (*L*/4.0).
3. Optimize the dimensions (*L*, *W*) of the RMSA to tune the resonant peak to f_r.
4. Further, optimize the slot dimensions L_v, L_h, and w_s so that the loop due to the coupling of TM_{10} and TM_{02} modes lies inside the VSWR $= 2$ circle as shown in Fig. 3.2a.

The above procedure is repeated for frequencies in the range of 800–5000 MHz at a step size of 200 MHz, for the corresponding different values of 'h' for the given U-slot RMSA.

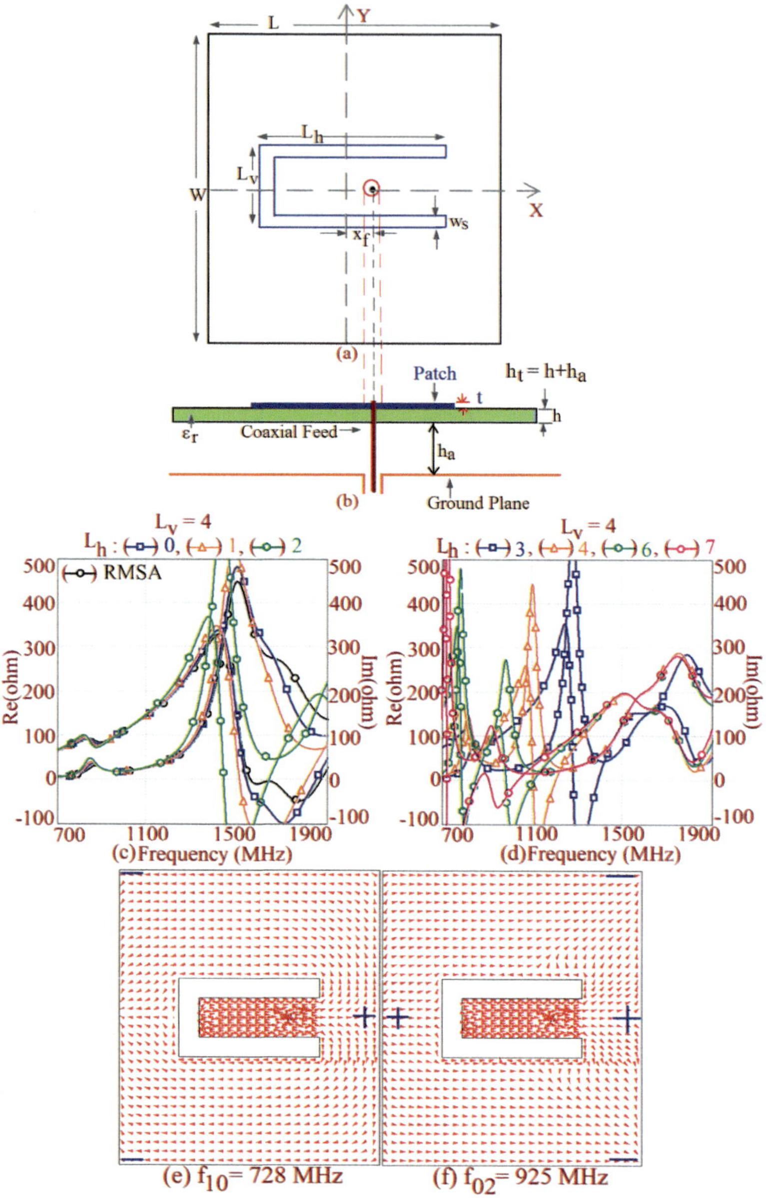

Fig. 3.1 **a**, **b** RMSA with U-slot, **c**, **d** resonance frequency plots for varying lengths, and its **e**, **f** surface current distributions at U-slot cut patch modes [26]

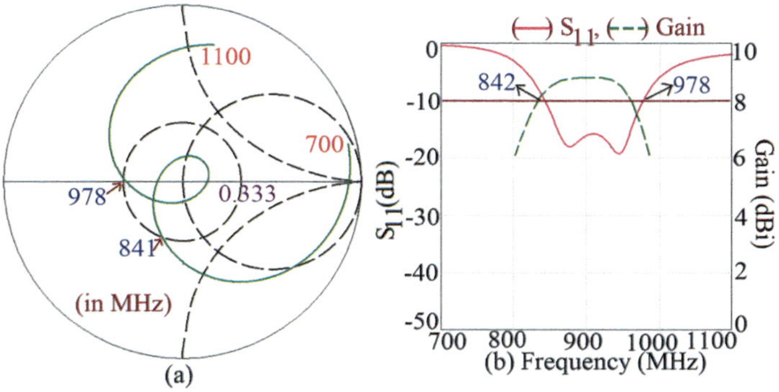

Fig. 3.2 Simulated **a** smith chart, and **b** S_{11}, gain plots of RMSA with U-slot [26]

The ANN model for U-slot cut RMSA is designed using the feed-forward supervised networks in two different designs, i.e., single-stage ANN model and double-stage ANN model, which are explained below. In both the models, the ANN model is proposed for varying substrate thicknesses (h) over a wide frequency range of 800–5000 MHz. At all the frequencies, while predicting the resonance frequency (patch length), the substrate thickness is taken in the range of 0.05λ–0.1λ.

3.2.1.1 Single-Stage ANN Model

The input parameters applied to the neural network are the resonance frequency of RMSA and substrate thickness (h) of the microstrip antenna. The (ΔL)/(h) and the U-slot parameters (horizontal slot length, vertical slot length), patch thickness, and the feed point location were given as the target data. Using (ΔL)/h, the patch length (L) of the rectangular MSA was obtained. By maintaining a ratio of 5:6 (length: width), the width (W) was obtained from the patch length. The neural network was then trained using the above-mentioned data for a frequency range of 800–5000 MHz. Four different substrate thicknesses in the range of 0.05λ–0.1λ were applied to each frequency. In total, 84 data points were fed to the neural network model. After training the network, it was tested on different frequencies for different substrate thicknesses. The block diagram of the single-stage ANN model and the flowchart showing the data creation and training process are shown in Fig. 3.3a, b. The sample training data set of single-stage ANN model created for the frequency range of 800–5000 MHz on varying substrate thickness forms 0.05λ–0.1λ is shown in Table 3.1.

Further, the trained ANN structure is used to predict the dimensions of U-slot cut RMSA for a given TM_{10} mode frequency (f_r) and substrate thickness (h). The dimensions predicted by the single-stage ANN architecture is given in Table 3.2, after minor changes. These minor changes in the dimensions are done to obtain practically realizable values.

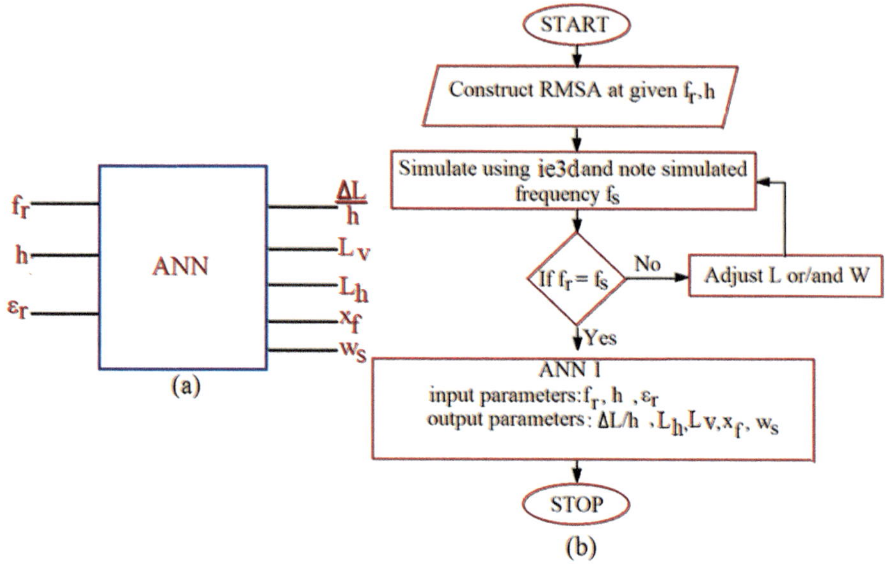

Fig. 3.3 **a** Block diagram, **b** flowchart of single-stage ANN model

Table 3.1 ANN training data set for U-slot MSA

f_r	f_{ie3d}	h	$\Delta L/h$	L	W	L_h	L_v	x_f	w_s
800	805.6	2.25	0.08044	15	18	9.2	5.3	1.2	0.9
1000	1005.6	1.8	0.08101	12	14.4	6.85	4.15	1.5	0.7
1200	1194.4	2	0.07397	9.6	11.52	5.8	3.6	0.6	0.6
1400	1404.8	1.1	0.08535	8.8	10.56	5.4	3.1	1.3	0.55
1600	1600	1.5	0.0725	7.2	8.64	4.4	3.2	0.3	0.45
1800	1811.2	0.833	0.08894	6.8	8.16	3.2	1.95	1.5	0.4
2000	2020	0.9	0.08476	5.9	7.08	3.6	2.3	0.7	0.4
2200	2214.88	0.682	0.08595	5.6	6.72	3.05	1.8	0.85	0.35
2400	2382.48	0.875	0.07977	4.9	5.88	2.9	1.7	0.5	0.3
2800	2795.04	0.75	0.08111	4.15	4.98	2.3	1.5	0.5	0.25
3000	2982.48	0.5	0.08794	4.15	4.98	2.15	1.25	0.75	0.25
3200	3212	0.75	0.0808	3.458	4.15	2.08	1.4	0.2	0.2
3400	3400	0.441	0.09396	3.583	4.3	2	1	0.65	0.2
3600	3600	0.667	0.08123	3.083	3.7	2.14	1.4	-0.05	0.2
4000	4072	0.6	0.07364	2.8	3.36	2	1.2	-0.1	0.2
4400	4400	0.34	0.0881	2.81	3.372	1.35	0.75	0.6	0.15
4600	4600	0.456	0.07795	2.55	3.06	1.5	1.1	0.2	0.15
4800	4772.56	0.5	0.0743	2.4	2.88	1.35	1	0.2	0.15
5000	5034	0.3	0.08829	2.45	2.94	1.16	0.68	0.53	0.15

Table 3.2 Optimized and practically realizable ANN values of U-slot MSA

f_r	h	$\Delta L/h$	L	W	L_h	L_v	x_f	w_s
600	2.5	0.72193754	21.39	25.668	9.528	5.397	2	0.863
700	2.15	0.80815321	17.954	21.545	9.487	5.388	0.887	0.895
900	1.7	0.86820924	13.715	16.458	8.428	3.158	1.995	0.809
1500	1.4	0.77831323	7.821	9.385	4.497	3.842	0.454	0.517
2100	0.75	0.84573274	5.874	7.049	3.34	2.174	0.589	0.399
2700	0.8	0.82084014	4.242	5.09	2.462	1.538	0.403	0.255
3300	0.5	0.84199431	3.703	4.444	2.039	1.233	0.451	0.213
3900	0.4	0.83706793	3.176	3.811	1.836	1.155	0.242	0.18
4300	0.5	0.79565248	2.693	3.232	1.577	1.02	0.325	0.175
4900	0.35	0.80299333	2.499	2.999	1.294	0.862	0.276	0.151
5100	0.4	0.75819635	2.335	2.802	1.16	0.694	1.989	0.154
5500	0.3	0.7219	2.294	2.753	9.6	5.203	-0.053	0.15
5900	0.4	0.7219	1.965	2.358	9.6	5.396	-0.1	0.189

The transgression graph for single-stage ANN using the mean squared error function is plotted against the number of epochs used during the training algorithm. The results as displayed below depict the Validation, Target, and Test Results. Mean squared error function is used to calculate the average squared difference between the targets and the outputs. As observed, the lower the value of MSE, the more effective is the neural network. The performance plot for the model is shown in Fig. 3.4a. The regression graph is a data plot that graphs the linear relationship between the independent and dependent variables. It is typically used to visually show the strength of the relationship and the dispersion of results, all to explain the behaviour of the dependent variable, which is shown in Fig. 3.4b–e.

3.2.1.2 Double-Stage ANN Model

The performance of the ANN in terms of MSE can be improved by increasing the number of epochs or by increasing the number of layers in the ANN model. In the double-stage design of ANN model, two layers of neural networks are used, where the output of the first layer is given as an input to the second layer network. The first neural network layer is modelled to predict the patch dimensions of the RMSA. So, the inputs given to the first network are frequency and substrate thickness. The target data for this network is $(\Delta L)/h$. Thus, the value of $(\Delta L)/h$ is predicted for required frequencies and substrate thicknesses using this network. Here the length and width of the RMSA are calculated using $(\Delta L)/h$, and maintaining $L/W = 5{:}6$, as observed from the initial optimum configuration designed at 900 MHz. Since the patch dimensions of the MSA are obtained, these dimensions along with the frequency and substrate thickness are given as input to the second neural network.

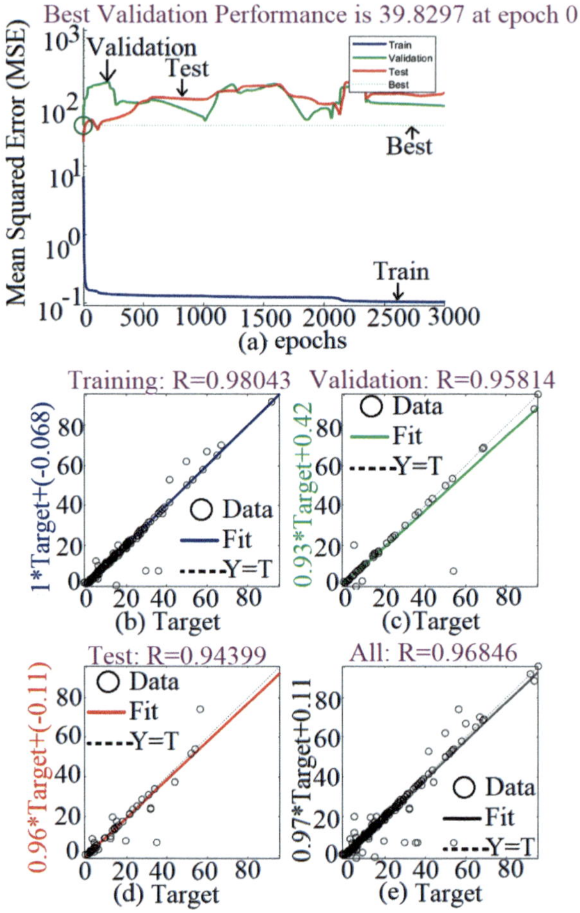

Fig. 3.4 **a** Performance graph, **b–e** regression graphs of predicted values of single-stage ANN model of U-slot MSA

The target data given is horizontal slot length, vertical slot length, patch thickness, and the feed point location. After training the second neural network on this data, the U-slot parameters are predicted. The result of this two-step ANN is very efficient with the requirement of very fine optimization in some of the U-slot parameters. Also, some defined optimization is done to generate practically realizable values of the MSA. The above procedure is repeated for frequencies in the range of 800–5000 MHz at a step size of 200 MHz, for corresponding different values of 'h_t' for the given U-slot cut RMSA.

For the training of the double-stage ANN model, same data set which is used for single-stage ANN model is considered, as given in Table 3.1. The flowchart for training double-stage U-slot cut RMSA ANN model is shown in Fig. 3.5. Also, the block diagram and ANN model of double-stage ANN model are shown in Fig. 3.6a, b respectively. The neural network is based on the feed-forward backpropagation network algorithm that comprises the three layers each. Two fully connected neural

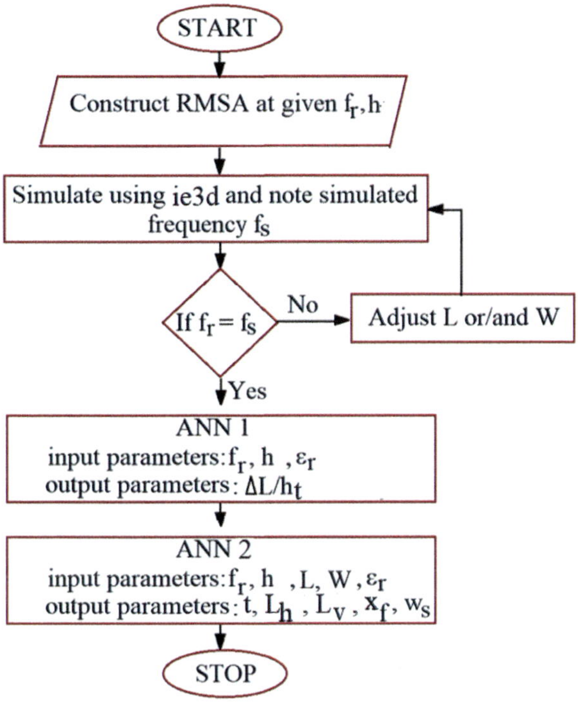

Fig. 3.5 Flowchart of training process of U-slot cut RMSA using double-stage ANN model

network models are made up of input units (green), hidden units (Gray), and output units (blue), with all the units connected to all the units in the layers, on either side. Inputs are fed in from the left, then activate the hidden units in the middle, and make outputs feed out from the right. The strength (weight) of the connection between any two units is gradually adjusted as the network learns. The parameters of the first-layer ANN and second-layer ANN in the double-stage model are given in Table 3.3.

The number of inputs, outputs, and hidden neurons are fed to the neural network in accordance with the training needed. The number of epochs is set as high as possible to terminate the training based on the error rates generated. The LM algorithm, also known as the DLS method, is used to solve nonlinear least square problems. These minimization problems arise, especially in least square curve fitting. The data set is automatically divided into training, validation, and test sets when the network is ready to be trained. The training set is used to teach the network. Training continues as long as the network continues improving on the validation set. The test set provides a completely independent measure of network accuracy. To see the network's performance during training, performance graphs are plotted which are shown in Fig. 3.7a, b. Performance is measured in terms of mean squared error and shown on a log scale. It rapidly decreases as the network is trained. Performance is shown for each of the training, validation, and test sets. The final network is the network that performed best on the validation set. A training set is presented to the network during training, and the network is adjusted according to its error. A validation set is used to measure network

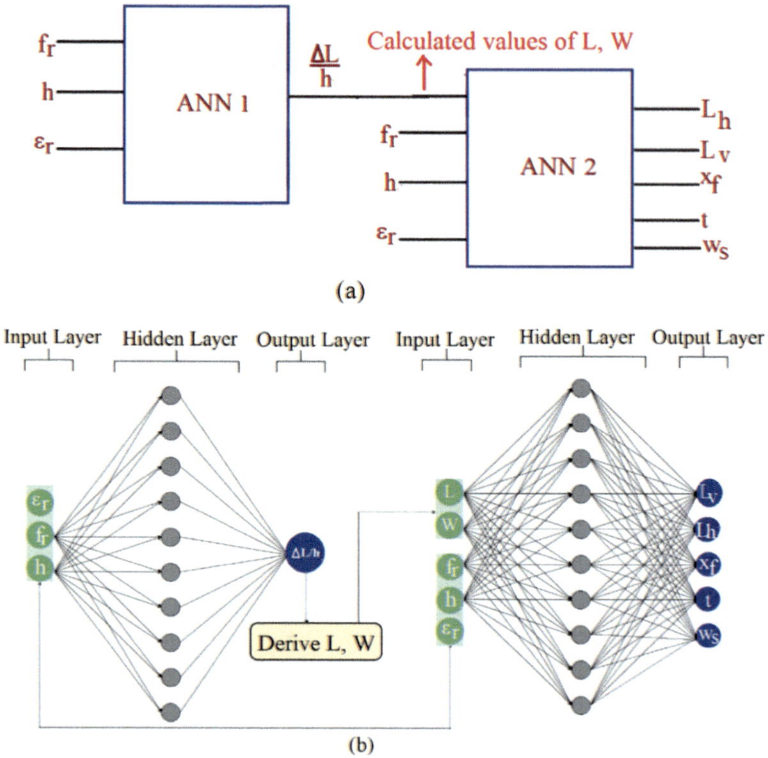

Fig. 3.6 **a** Block diagram, **b** ANN model with all layers of double-stage ANN model for U-slot cut RMSA

Table 3.3 Model parameters in both stages of two-step ANN model

Sr. No.	Parameter	Values in first stage	Values in second stage
1	Number of inputs	3	5
2	Number of outputs	1	4
3	Number of hidden neurons	10	10
4	Number of epochs	186	2000
5	Training algorithm	LM	LM

generalization and to halt training when generalization starts improving. Test data does not affect training and so provides an independent measure of network performance during and after training. The Transgression graph for double-stage ANN using MSE function is plotted against the number of epochs used during the training algorithm. The results as displayed in Fig. 3.7 depict the Validation, Target, and Test Results. MSE function is used to calculate the average squared difference between the targets and the outputs. As observed, the lower the value of MSE, the more effective is the neural network. Compared to the single-stage ANN network, the performance graph of the double-stage ANN network is observed to be exponentially increasing. The mean squared error is reduced by at least 450 units which increases the accuracy of the prediction. Another measure of how well the neural network has fit the data is the regression plot. Here the regression is plotted across all samples as shown in Fig. 3.7c–f. The regression plot shows the actual network outputs plotted in terms of the associated target values. If the network has learned to fit the data well, the linear fit to this output-target relationship should closely intersect the bottom-left and top-right corners of the plot. If this is not the case then further training, or training a network with more hidden neurons, is done.

Further, the trained ANN structure is used to predict the dimensions of U-slot cut RMSA for a given TM_{10} mode frequency (f_r) and substrate thickness (h). The dimensions predicted by the single-stage ANN architecture are given in Table 3.4, after minor changes. These minor changes in the dimensions are done to obtain practically realizable values.

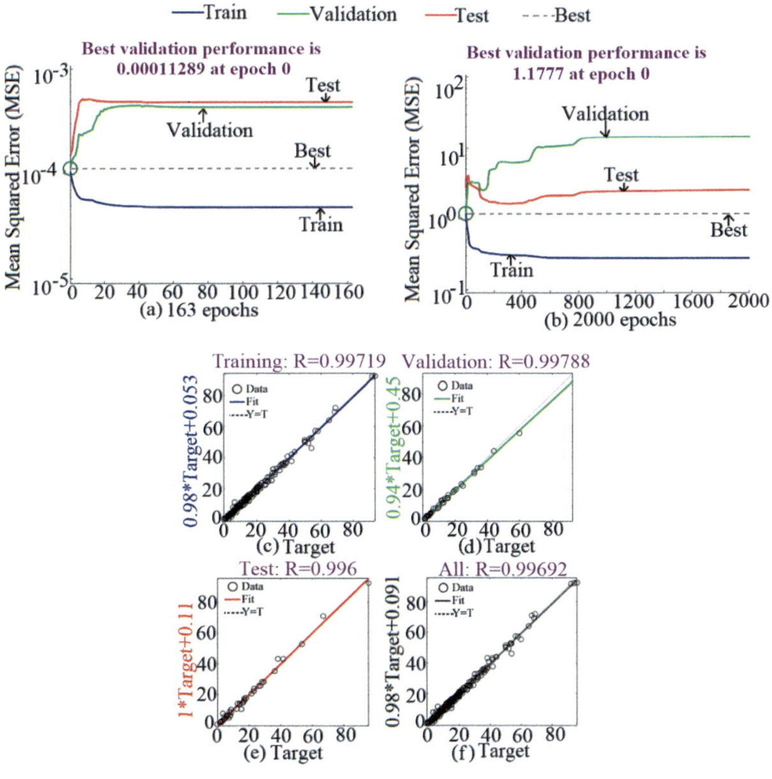

Fig. 3.7 **a, b** Performance graphs, **c–f** regression graphs of predicted values of double-stage ANN model of U-slot MSA

Table 3.4 Predicted dimensions of U-slot MSA using double-stage ANN model

f_r	h	$\Delta L/h$	L	W	L_h	L_v	x_f	w_s
600	25	0.95	202.50	243.00	95.85	53.97	19.39	8.99
600	35	0.7219	199.47	239.36	95.96	54.00	8.29	8.98
700	21.5	0.97045	172.56	207.07	95.05	53.80	17.93	8.95
700	30	0.721923	170.97	205.16	95.56	53.93	9.47	8.89
800	19	0.952	16.89	8.64	151.32	181.58	89.53	52.14
900	17	0.965843	133.83	160.60	76.37	46.78	16.46	7.88
900	23.5	0.75465	131.20	157.44	79.61	48.67	11.54	7.43
1200	12.5	0.869	18.99	6.72	103.26	123.91	51.77	30.23
1500	10	0.866619	82.67	99.20	38.73	27.06	15.25	4.83
1500	14	0.778498	78.20	93.84	46.71	35.99	1.90	4.35
1600	3	0.791	4.27	4.26	73.17	87.8	42.6	31.18
1700	12	0.989	3.86	4.3	64.48	77.38	40.37	31.98
2100	7.5	0.893239	58.03	69.64	34.70	24.34	4.80	3.70
2100	10	0.790632	55.62	66.74	33.59	21.84	4.85	3.51
2700	6	0.846838	45.39	54.47	24.20	13.76	7.97	2.78
2700	8	0.79491	42.84	51.41	25.01	15.13	4.12	2.58
3300	5	0.912051	36.33	43.60	19.44	11.66	5.23	2.10
3300	6.5	0.831672	34.64	41.57	20.94	13.08	1.90	1.94
3900	4	0.834366	31.79	38.15	17.09	10.75	4.92	1.89
3900	5.5	0.751719	30.19	36.23	19.71	12.92	0.99	1.75
4300	4	0.831768	28.23	33.88	15.16	9.54	3.62	1.72
4300	5	0.783283	27.05	32.46	16.32	10.51	1.27	1.67
4900	4	0.781076	24.36	29.23	13.26	8.11	3.99	1.62
5100	3	0.853454	24.29	29.15	12.78	7.76	5.72	1.62
5500	3	0.761174	22.71	27.25	12.51	7.56	5.94	1.60
5900	3	0.75041	20.92	25.10	12.38	7.47	6.10	1.59

From the predicted data of U-slot cut RMSA, dimensions for the design at 800 MHz are considered and are selected further to get the practically realizable values. The dimensions of U-slot cut RMSA are $L = 15.1$, $W = 18.2$, $h = 1.9$, $L_h = 8.5$, $L_v = 5.2$, $w_s = 0.9$, and $x_f = 1.7$ cm. The simulated and measured S_{11} BW and gain plots are shown in Fig. 3.8a. A simulated BW of 147 MHz (17%) and measured BW of 140 MHz (16.27%) with a peak gain of 9.27 dBi are realized. The fabricated prototype of U-slot cut RMSA at 800 MHz on a substrate thickness of $0.05\lambda_g$ is shown in Fig. 3.8b, c.

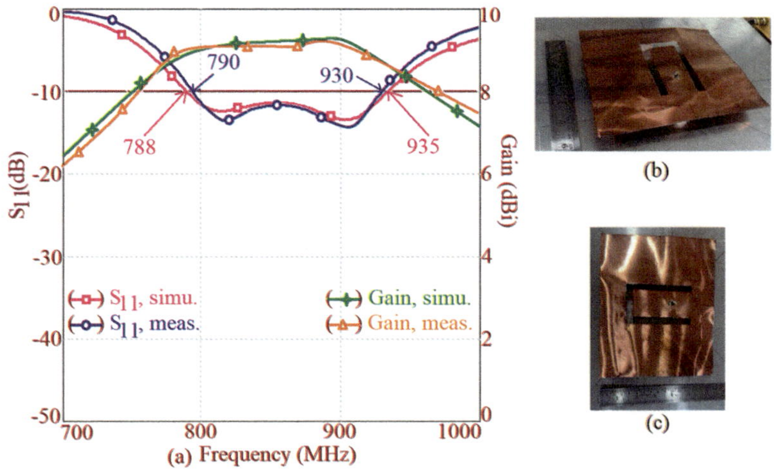

Fig. 3.8 **a** S_{11} BW and gain plots, and **b**, **c** fabricated prototype of U-slot MSA designed at 800 MHz using double-stage ANN model

3.2.2 U-Slot Cut RMSA with Asymmetric Slot Lengths for CP Response

The U-slot cut RMSA discussed in the previous section realizes wideband response due to the optimum spacing between TM_{10} and TM_{02} resonant modes against the optimization of horizontal and vertical U-slot lengths. This U-slot cut RMSA exhibits linear polarization over the entire S_{11} BW [26]. However, in some applications such as mobile communication, CP response is preferred to avoid signal loss. To achieve CP, orthogonal resonant modes are to be realized, and one of the methods of realizing this is creating symmetry in the structure of the slot cut MSA [27]. One such configuration of SMSA with asymmetric U-slot reported in [28] is shown in Fig. 3.9a, b.

As described in [28], an increase in L_v reduces the TM_{10} mode frequency as the slot length is orthogonal to the modal surface currents of TM_{10} mode. Further, with an increase in L_h, the frequency of TM_{02} mode reduces, but TM_{10} mode reduction is marginal as shown in Fig. 3.9c [28]. To create asymmetry in the structure, one of the horizontal arms of the U-slot is further extended. With increase in L_{h1}, less reduction in TM_{02} mode was observed. However, TM_{10} mode gets degenerated into ${TM_{10}}^{45}$, and ${TM_{10}}^{135}$ near TM_{10} mode frequency only [28]. With an increase in l_{h1}, the frequency of the ${TM_{10}}^{45}$ mode reduces whereas that of the ${TM_{10}}^{135}$ remains constant. Across the two degenerated modes, surface current vectors are orthogonal and the optimum separation between the two modes yields a CP response for AR less than 3 dB, as shown in Fig. 3.9d. The orthogonal surface current distributions of the degenerated modes are shown in Fig. 3.9e, f [28]. The next resonant mode in unequal length U-slot SMSA corresponds to the modified TM_{02} mode, and the

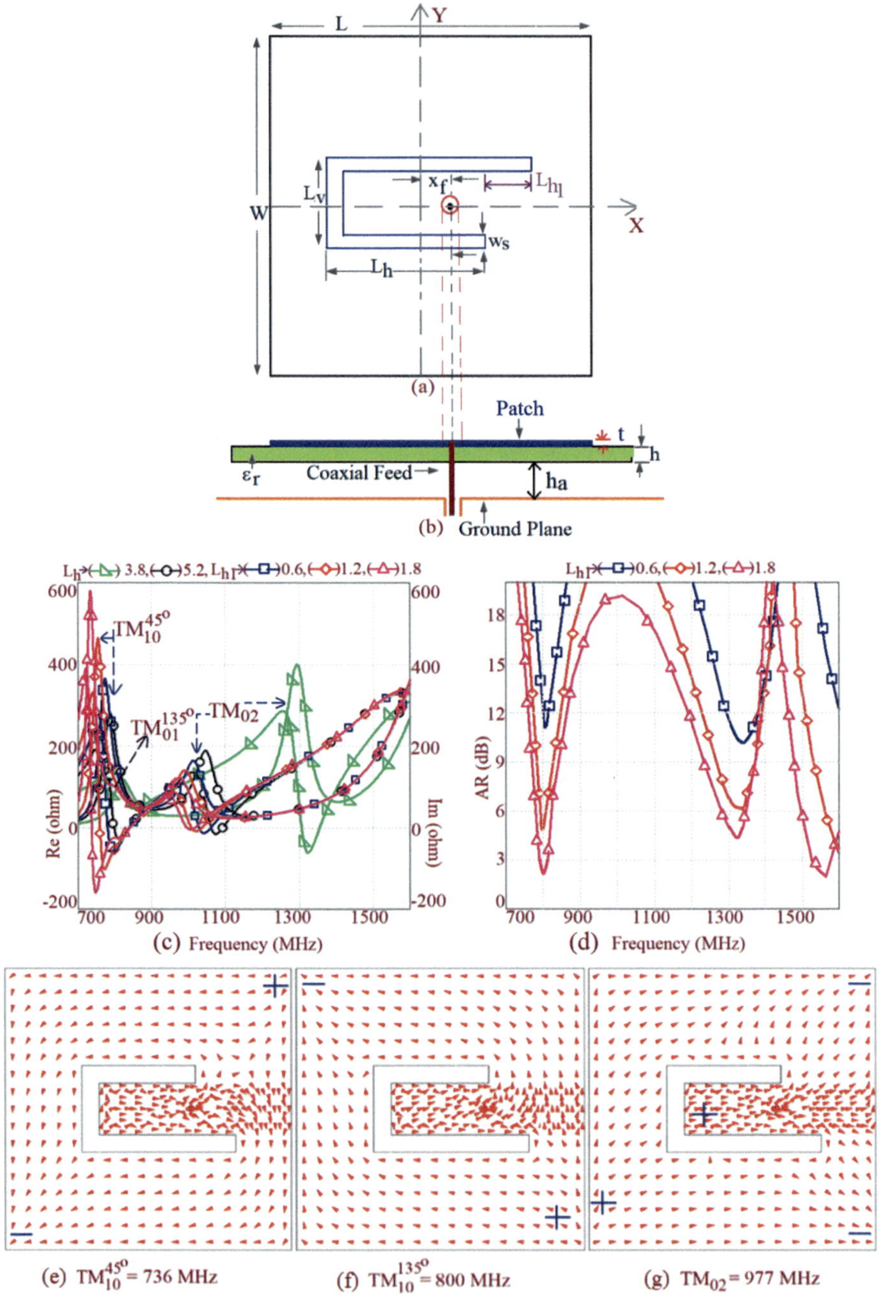

Fig. 3.9 **a**, **b** Asymmetric U-slot cut square MSA, its **c** resonance frequency, and **d** AR plots for varying slot dimensions, **e**–**g** surface current distributions for Asymmetric U-slot cut SMSA [28]

surface currents at the same are modified due to the presence of unequal length U-slot as shown in Fig. 3.9g [28].

An optimum CP response in unequal length U-slot cut SMSA is obtained for $L_v = 4.2$, $L_h = 5.1$, $L_{h1} = 1.8$, and $x_f = 2.2$ cm, and results for the same are shown in Fig. 3.10a, b. The U-slot cut RMSA with unequal lengths yields a dual band response with the CP response in the first band [28]. The first band is due to the coupling between the dual degenerated TM_{10} modes, and the second band is due to the modified TM_{02} mode. The simulated values of the impedance BW in the two bands are 107 MHz (12.9%) and 54 MHz (5.27%). Respective measured values are 112 MHz (13.64%) and 60 MHz (5.81%) [28]. The simulated and measured AR BW in the first band is 27 MHz (3.37%) and 25 MHz (3.16%), respectively [28]. In the two bands, broadside peak gain is above 8 dBi and offers stable gain characteristics [28]. In both the bands, the antenna exhibits a broadside radiation pattern showing equal contribution of the co- and cross-polar levels, due to the CP response. The right-hand fields are dominant in the radiated signal, indicating the presence of RHCP wave.

Further, to design ANN model for asymmetric U-slot SMSA for CP response, the crucial step is in generating training data set. For that, initially the SMSA is designed on an air-suspended substrate with total thickness $h_t = h_a + h$, for its fundamental TM_{10} mode over a frequency range of 800–3000 MHz and in a step size of 200 MHz. At each frequency, asymmetrical U-slot cut SMSA is optimized for substrate heights, varying in the range $0.05\lambda_g$–$0.1\lambda_g$. The MSA design is carried out in the following manner for each frequency and for each substrate thickness to create training data set for the ANN, and the detailed method of data set creation is represented in the flowchart as shown in Fig. 3.11.

1. Design SMSA for a particular TM_{10} mode frequency f_r, and total substrate thickness h_t using resonant frequency equation of RMSA [2].

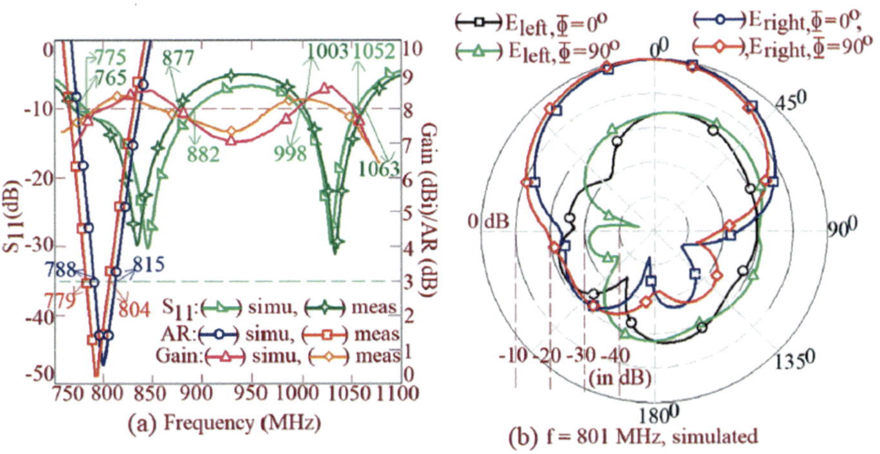

Fig. 3.10 **a** Simulated and measured results, and **b** field polarization radiation pattern plots for asymmetric U-slot cut SMSA for CP response in the first band [28]

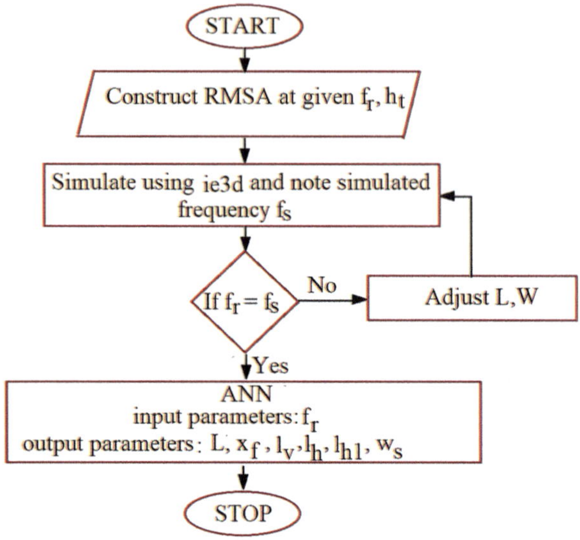

Fig. 3.11 Flowchart of training process of asymmetric U-slot cut SMSA

2. Feed the SMSA at appropriate location starting at (*L*/4,0).
3. Optimize the length '*L*' of the SMSA to tune the resonant peak to f_r.
4. Further, optimize the slot dimensions L_v, L_h, L_{h1}, w_s, and feed point location x_f so that two bands are observed in the S_{11} plot, wherein CP response is realized in the first band due to orthogonal $TM_{10}{}^{45}$, $TM_{10}{}^{135}$ modes and second band is due to the modified TM_{02} modes. From the optimum configurations designed at various frequencies, the relations between various parameters and patch dimensions are selected such as '$L_h = 0.302\ L$', '$L_{h1} = 0.53\ L$', and '$L_v = 0.323\ L$'.

For the training purpose of the neural network model, simulations were carried out using IE3D software [29] over a frequency range of 800–3000 MHz in a step size of 200 MHz. For each frequency, antenna designs were simulated for four different substrate heights, i.e., $0.05\lambda_g$–$0.08\lambda_g$. In total there were 84 points of data to train the neural network model. The sample training data set created using the above procedure is given in Table 3.5. The block diagram and detailed structure of the ANN model for the trained data are shown in Fig. 3.12a, b. A four-layer network is used in the ANN model of asymmetric U-slot cut SMSA. The first layer is the input layer, which has only one neuron that has frequency as one of its properties. There are two hidden layers, one with 6 neurons with ReLU activation function to provide nonlinearity to the model and the other with all 8 neurons as outputs for asymmetric U-slot cut antenna with sigmoid activation function. To prevent a slow learning gradient descent curve for the training model, MinMaxScalar is used to scale the features. Because the generated data set has no outliers, mean squared error is utilized as the loss function. Adam optimizer is used in the ANN because of its twin benefits of stochastic gradient descent with momentum and adaptive learning of the learning rate. The details of the ANN architecture such as number of inputs, outputs, and hidden layers are specified in Table 3.6.

Using the proposed ANN model which got trained with the above training data set, antenna dimensions are predicted for a frequency range of 800–3000 MHz in steps of 200 MHz. The dimensions predicted by the ANN architecture are given in Table 3.7, after minor changes. These minor changes in the dimensions are done to obtain practically realizable values. The mean squared errors of trained and predicted dimensions by ANN with varying epochs are shown in Fig. 3.13a. A reduction in

Table 3.5 ANN training data set for asymmetric U-slot MSA

F_R	λ_G	H_T	L	L_H	L_{H1}	L_V	W_S	X_F
700	41.82429	2.718579	15.60602	6.117561	2.153631	5.040745	1.200103	2.395524
900	32.53	2.11445	12.13802	4.758103	1.675046	3.92058	0.933414	1.863186
1100	26.61546	1.730005	9.931105	3.892993	1.370492	3.207747	0.763702	1.524425
1300	22.52077	1.46385	8.403243	3.294071	1.159647	2.714247	0.646209	1.289898
1500	19.518	1.26867	7.28281	2.854862	1.005028	2.352348	0.560048	1.117911
1700	17.22177	1.119415	6.426009	2.518996	0.886789	2.075601	0.49416	0.986392
1900	15.40895	1.001582	5.749587	2.253838	0.793443	1.857117	0.442143	0.882562
2100	13.94143	0.906193	5.202007	2.039187	0.717877	1.680248	0.400034	0.798508
2300	12.72913	0.827394	4.749659	1.861866	0.655453	1.53414	0.365249	0.729073
2500	11.7108	0.761202	4.369686	1.712917	0.603017	1.411409	0.336029	0.670747
2700	10.84333	0.704817	4.046006	1.586034	0.558349	1.30686	0.311138	0.621062
2900	10.09552	0.656209	3.766971	1.476653	0.519842	1.216732	0.28968	0.57823

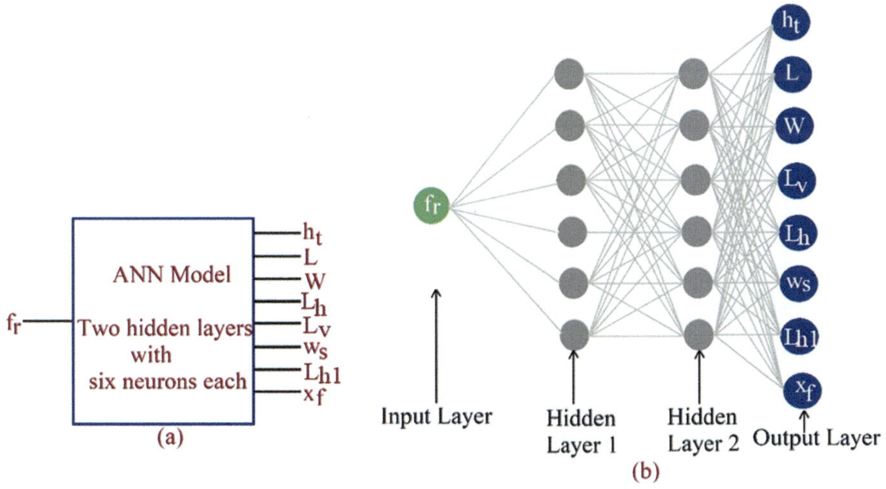

Fig. 3.12 **a** Block diagram, **b** ANN model of Asymmetric U-slot MSA

Table 3.6 Number of inputs and outputs in ANN

Number of inputs	1
Number of outputs	8
Number of hidden nodes	12
Number of hidden layers	2
Number of hidden nodes per layer	12/2 = 06
Number of epochs	1500
Training algorithm	ANN

error is observed with a large number of epochs. From the dimensions observed from the table, the asymmetric U-slot cut SMSA is designed at 1000 MHz at $h_t = 0.065\lambda_g$, and simulations are conducted to verify the result in terms of dual band and CP response. All the simulated and measured results are shown in Fig. 3.13b, and the fabricated prototype of asymmetric U-slot cut SMSA designed at 1000 MHz from ANN predicted dimensions is shown in Fig. 3.13c. The simulated values of the impedance BW in the two bands are 147 MHz (15%) and 82 MHz (6.4%). Respective measured values are 126 MHz (12.92%) and 80 MHz (4.1%). The simulated and measured AR BW in the first band are 16 MHz (1.675%) and 16 MHz (1.67%), respectively. In the two bands, broadside peak gain is about 8.54 dBi and offers stable gain characteristics across the impedance and AR BW. In both the bands, the antenna exhibits a broadside radiation pattern showing equal contribution of co- and cross-polar levels, due to the CP response.

Table 3.7 ANN predicted and optimized dimensions of asymmetric U-slot MSA

F_R	λ_G	H_T	L	L_H	L_{H1}	L_V	W_S	X_F
800	391.468	25	137	54.5	22	47	11	22
1000	313.17	20	106	43	18	38	9	17.5
1200	193.48	12.6	86	33	10	23	6	11.2
1400	162.54	10.6	72	30	7	20	5	9.5
1600	154.45	10.6	60	24	9	19	4.8	9.5
1800	147.62	10	55	22.5	8	16	4	9.5
2000	141.93	9.6	48	20	7	14	4	8.5
2200	137.214	9	42	17	8	15	3.5	8.5
2400	133.35	8	41	17.5	7	13	3.5	8
2600	130.19	7.6	36	14	6	14	3	8
2800	127.62	7	32	14.5	4	10	2.5	6
3000	125.54	6	32	12.6	5	9.2	2.4	5.5

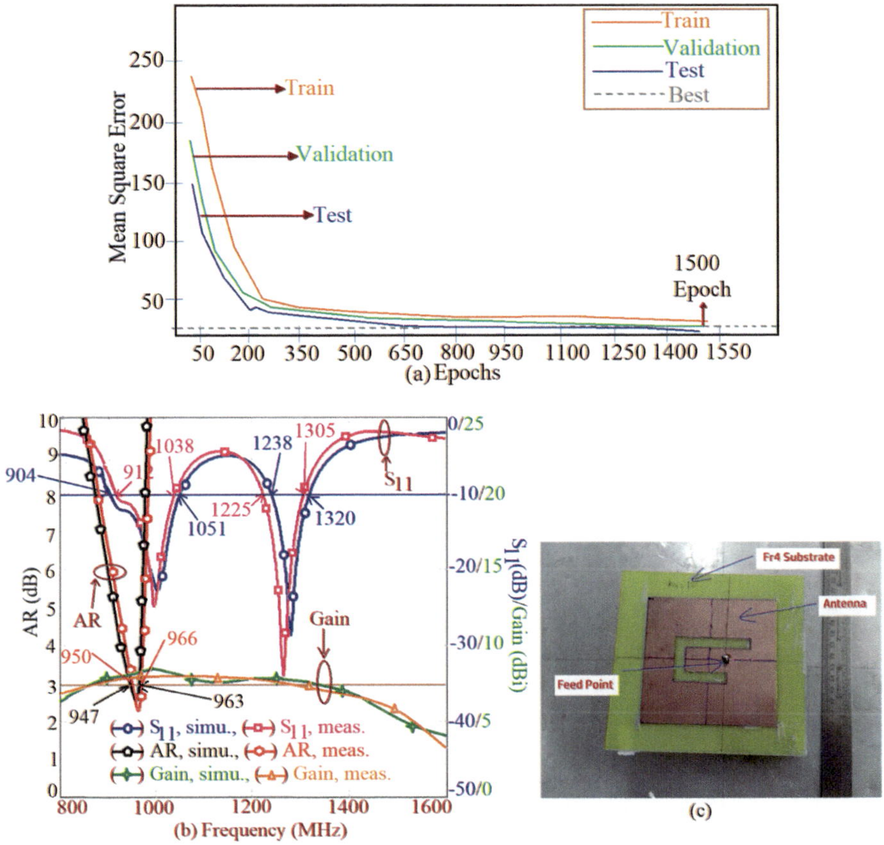

Fig. 3.13 **a** MSE versus epochs graph, **b** simulated and measured results, and **c** fabricated prototype of asymmetric U-slot cut SMSA designed from predicted dimensions using ANN model at 1000 MHz on $0.065\lambda_g$ thickness

3.2.3 ANN Model of Truncated SMSA with U-Slot for Dual Band and CP Response

As discussed in the previous section, dual band CP response can be realized by creating asymmetry in SMSA in terms of asymmetrical U-slot length. Another variation of SMSA with symmetrical U-slot, i.e., equal lengths for horizontal sections in U-slot, which realizes dual band CP response is reported in [28]. Here, the asymmetry in SMSA is created by truncating its corners as shown in Fig. 3.14a, b. On SMSA, only one resonant mode is excited with respect to the specified feed point location. With the introduction of truncation and tuning of its length '*d*', the single resonant mode gets degenerated into TM_{10} and TM_{01} modes. Further, with the increment in the U-slot dimensions, TM_{10} mode frequency reduces due the circulation of surface currents around the U-slot. Reduction in TM_{02} mode is also observed as the horizontal slot length is in perpendicular direction to that of the surface currents at this mode, while TM_{01} mode is marginally affected due to the minimal change in the corresponding surface current directions. The variations in different modes with U-slot dimensions are shown in Fig. 3.14c, d. The resonant mode-based working of this corner-truncated antenna is explained in detail in [28].

The U-slot length realizes tuning in between the orthogonal TM_{10} and TM_{01} mode frequencies and yields CP response for AR less than 3 dB, as shown in Fig. 3.15a. Thus, this design also yields CP response in the first band due to degenerated TM_{10}, TM_{01} modes and dual band response due to the modified TM_{02} mode. The simulated value of the impedance BW in the two bands is 97 MHz (11.6%) and 84 MHz (7.71%),

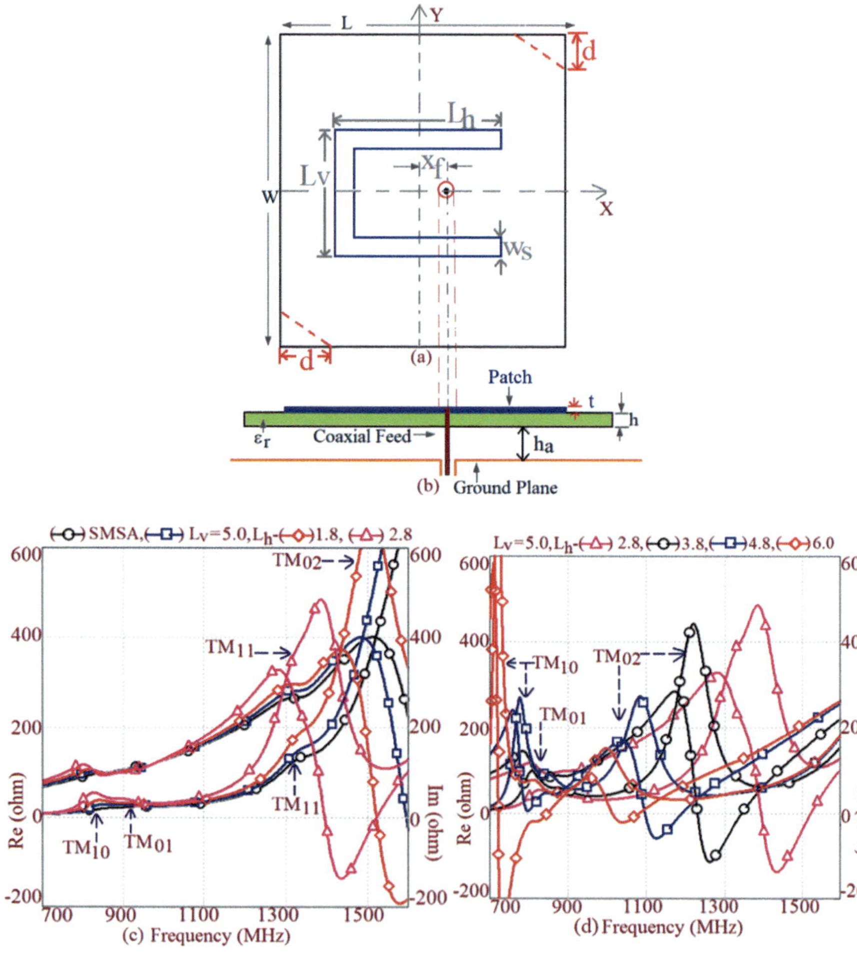

Fig. 3.14 **a**, **b** Corner-truncated SMSA with U-slot, its **c**, **d** simulated resonance frequency plots for U-slot length variations [28]

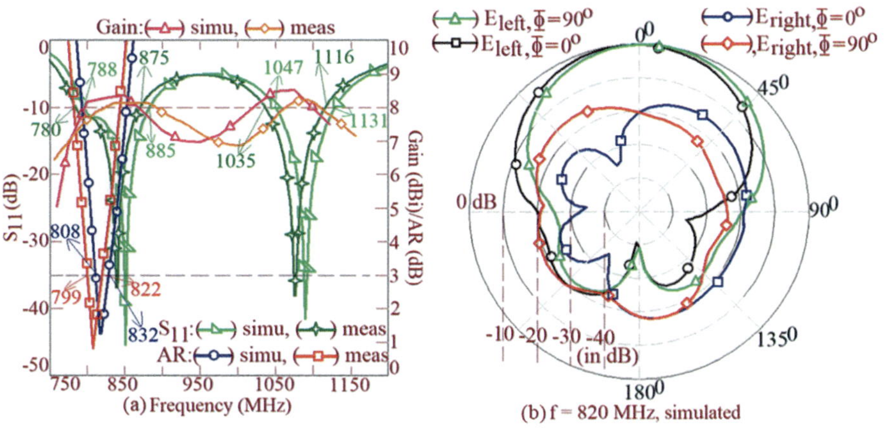

Fig. 3.15 **a** Simulated and measured results, and **b** field polarization plots for corner-truncated SMSA with U-slot [28]

respectively. Respective impedance BW observed in the measurement is 95 MHz (11.48%) and 81 MHz (7.53%). The AR BW as observed in the first band in the simulation and measurement is 24 MHz (2.93%) and 23 MHz (2.84%), respectively. For the mentioned corner truncation, MSA exhibits a left-hand CP (LHCP) wave in the first band as shown in Fig. 3.15b. The gain response is stable across the AR BW. In both bands, a broadside radiation pattern is observed over the impedance and AR BW with a peak gain of above 8 dBi.

Further, to design the ANN model for corner-truncated U-slot cut SMSA for CP response, the crucial step is generating training data set for the neural network model. The procedure for generating training data set is like that of asymmetrical U-slot cut SMSA described in the previous section. The MSA design is carried out in a similar manner for each frequency and for each substrate thickness, to create the training data set for the ANN, and the detailed method of data set creation is represented in the flowchart, shown in Fig. 3.11.

From the optimum configurations designed at various frequencies, the relations between various parameters and patch dimensions are chosen as '$L_h = 0.484\,L$', '$L_v = 0.384\,L$'. '$w_s = 0.0615\,L$', '$d = 0.0138\,L$', and '$x_f = 0.3\,L$', in the training purpose. For the neural network training, simulations were carried out using IE3D Software [30] over a frequency range of 800–3000 MHz and in step size of 200 MHz. For each frequency, antenna designs were simulated at four different substrate thicknesses in the range $0.05\lambda_g$–$0.08\lambda_g$. In total, 84 data training points to train the neural network model are generated. Sample training data set created using the above procedure is given in Table 3.8. The block diagram and detailed structure of the ANN model for the trained data are shown in Fig. 3.12a, b. A four-layer network is used in the ANN model of asymmetric U-slot cut SMSA. The first layer is the input layer, which has only one neuron that has frequency as one of its parameters. There are two hidden layers, one with 6 neurons with ReLU activation function to provide nonlinearity to the model and the other with all 8 neurons as outputs for asymmetric U-slot cut antenna with sigmoid activation function.

To prevent a slow learning gradient descent curve for the training model, MinMaxScalar is used to scale the features. Because the generated data set has no outliers, mean squared error is utilized as the loss function. Adam optimizer is used in the ANN because of its twin benefits of stochastic gradient descent with momentum and adaptive learning of learning rate. The details of the ANN architecture such as number of inputs, outputs, and hidden layers are specified in Table 3.6. Using the proposed ANN model which got trained with the above training data set, antenna dimensions are predicted for a frequency range of 800–3000 MHz in steps of 200 MHz. The dimensions predicted by the ANN architecture are given in Table 3.9, after minor changes. These minor changes in the dimensions are done so as to obtain practically realizable values for different corner-truncated square patch loaded with U-slot.

The mean squared error of trained and predicted dimensions by ANN with varying epochs are shown in Fig. 3.16a. A reduction in error is observed with a large number of epochs. From the dimensions observed from the table, corner-truncated U-slot cut SMSA is designed for 1600 MHz frequency range on total substrate thickness of $0.065\lambda_g$, and simulations and measurements are conducted to verify the result in terms of dual band and CP response. All the simulated and measured results are shown in Fig. 3.16b, and the fabricated prototype of corner-truncated U-slot cut SMSA designed at 1600 MHz from ANN predicted dimensions is shown in Fig. 3.16c. The simulated values of the impedance BW in the two bands are 176 MHz (11.48%) and 161 MHz (8.12%). Respective measured values are 149 MHz (9.7%) and 137 MHz (6.9%). The simulated and measured AR BW in the first band are 38 MHz (2.5%) and 31 MHz (2.06%), respectively. In the two bands, broadside peak gain is about 8.25 dBi and offers stable gain characteristics across the impedance and AR BW. In

Table 3.8 ANN training data set for corner-truncated U-slot SMSA

F_R	λ_G	H_T	L	D	L_H	L_V	W_S	X_F
700	41.82429	2.7	15.6	3.2	6.9	6	1	4.681807
900	32.53	2.1	12.1	2.4	5.2	4.4	0.7	3.641405
1100	26.61546	1.76	10	2	4	3.4	0.6	2.979331
1300	22.52077	1.4	8.4	1.7	3.4	3	0.5	2.520973
1500	19.518	1.26	7.2	1.5	3.1	2.8	0.45	2.184843
1700	17.22177	1.16	6.4	1.3	2.8	2.5	0.4	1.927803
1900	15.40895	1.06	5.7	1.1	2.6	2.2	0.35	1.724876
2100	13.94143	0.96	5.2	1.05	2.2	2	0.3	1.560602
2300	12.72913	0.86	4.7	0.9	2.1	1.7	0.3	1.424898
2500	11.7108	0.76	4.4	0.9	1.9	1.6	0.26	1.310906
2700	10.84333	0.76	4	0.8	1.7	1.4	0.24	1.213802
2900	10.09552	0.66	3.6	0.7	1.6	1.3	0.23	1.130091

Table 3.9 ANN predicted and optimized dimensions of corner-truncated U-slot SMSA

F_R	λ_G	H_T	L	D	L_H	L_V	W_S	X_F
800	36.596	2.36	13.6	2.7	6.6	5.25	0.84	4.1
1000	29.277	1.96	11	2.2	5.3	4.2	0.65	3.3
1200	24.397	1.56	9	1.85	4.4	3.5	0.55	2.7
1400	20.912	1.36	7.8	1.6	3.7	3	0.4	2.3
1600	18.298	1.16	6.8	1.4	3.3	2.6	0.4	2
1800	16.265	1	6	1.2	2.9	2.3	0.3	1.8
2000	14.638	0.96	5.4	1.1	2.6	2.1	0.3	1.6
2200	13.3	0.86	5	1	2.4	1.9	0.3	1.5
2400	12.19	0.76	4.5	0.9	2.2	1.75	0.275	1.365
2600	11.26	0.76	4.2	0.85	2	1.65	0.25	1.26
2800	10.45	0.66	4	0.75	1.9	1.5	0.24	1.1
3000	9.759	0.66	3.6	0.74	1.75	1.4	0.22	1.1

both the bands, the antenna exhibits a broadside radiation pattern. In the first band where CP response is present, radiation pattern shows higher cross-polar radiation with equal contribution of the co- and cross-polar levels.

3.3 Summary

ANN modelling of U-slot and modified U-slot cut SMSAs for broadband and CP response are presented. For each design, its functioning is explained with respect to its resonance mode behaviour, and then the ANN model details along with nodal details, algorithm used, and criterion of the ANN such as MSE are provided. For wideband RMSA with U-slot, two approaches, i.e., ANN with single-stage and double-stage are provided. As ANN with double-stage approach predicts dimensions of U-slot cut RMSA, which yields better results in terms of performance curve and regression graphs, asymmetrical U-slot and corner-truncated U-slot cut SMSAs are modelled using the double-stage ANN approach. In the modelling of all U-slot configurations, ReLU activation function and MinMaxScalar are used to scale the features. With the help of the ANN model, the dimensions of the U-slot and modified U-slot MSAs can be predicted at a given frequency and substrate thickness, which helps in reducing the time required to optimize U-slot and modified U-slot MSAs by simulating them individually. A slight modification in the ANN predicted dimensions may be required to round off the dimensions to the nearest integer to get practically realizable dimensions. In all the MSAs, ANN predicted dimensions and calculated dimensions have shown less than 5% variation.

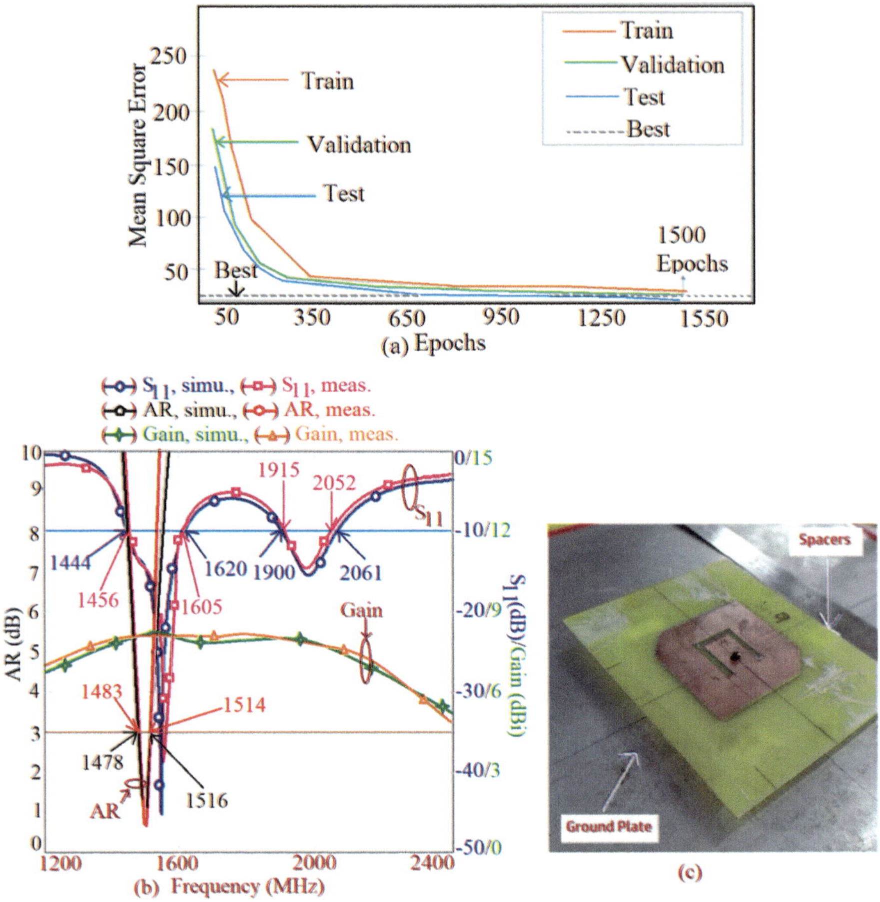

Fig. 3.16 **a** MSE versus epochs graph, **b** simulated and measured results, **c** fabricated prototype of corner-truncated U-slot SMSA designed from predicted dimensions at 1600 MHz on 0.065λg substrate thickness

References

1. Huynh T, Lee KF (1995) Single-layer single-patch wideband microstrip antenna. Electron Lett 31(16):1310–1312
2. Wong KL, Hsu WH (2001) A broadband rectangular patch antenna with a pair of wide slits. IEEE Trans Antennas Propag 49(9):1345–1347
3. Ansari JA, Yadav NP, Singh P, Mishra A (2011) Broadband rectangular microstrip antenna loaded with double U-slot slot. Int J Microwave Opt Technol Lett 6(4):185–190
4. Ge Y, Esselle KP, Bird TS (2006) A compact E-shaped patch antenna with corrugated wings. IEEE Trans Antennas Propag 54(8):2411–2413
5. Ooi BL, Shen Q (2005) A novel E-shaped broadband microstrip patch antenna. Microw Opt Technol Lett 27(5):348–352

6. Clenet M, Shafai L (1999) Multiple resonances and polarization of U-slot patch antenna. Electron Lett 35(2):101–102
7. Tong KF, Luk KM, Lee KF, Lee RQ (2000) A broadband U-slot rectangular patch antenna on a microwave substrate. IEEE Trans Antennas Propag 48(6):954–960
8. Lee KF, Yang SLS, Kishk AA, Luk KM (2010) The versatile U-slot patch antenna. IEEE Antennas Propag Magazine 52(1):71–88
9. Luk KM, Lee KF, Tam WM (1997) Circular U-slot patch with dielectric superstrate. Electron Lett 33(12):1001–1002
10. Luk KM, Guo YX, Lee KF, Chow YL (1998) L-probe proximity fed U-slot patch antenna. Electron Lett 34(19):1806–1807
11. Khodaei GF, Nourinia J, Ghobadi C (2008) A practical miniaturized U-slot patch antenna with enhanced bandwidth. Progress Electromagn Res B 3:47–62
12. Guo YX, Luk KM, Lee KF, Chow YL (1998) Double U-slot rectangular patch antenna. Electron Lett 34(19):1805–1806
13. Vedaprabhu B, Vinoy KJ (2010) An integrated wideband multifunctional antenna using a microstrip patch with two U-slots. Progress Electromagn Res B 22:221–235
14. Radavaram S, Pour M (2019) Wideband radiation reconfigurable microstrip patch antenna loaded with two inverted U-slots. IEEE Trans Antennas Propag 67(3):1501–1508
15. Wong KL (2002) Compact and broadband microstrip antennas. John Wiley & sons Inc., New York, USA
16. Bhardwaj S, Samii YR (2012) A comparative study of C-shaped, E-shaped, and U-slotted patch antennas. Microw Opt Technol Lett 54(7):1746–1756
17. Weigand SG, Huff H, Pan KH, Bernhard JT (2003) Analysis and design of broadband single layer rectangular U-slot microstrip patch antenna. IEEE Trans Antennas Propag 51(3):457–468
18. Ghalibafan J, Attari AR (2010) A new dual-band microstrip antenna with U-slotd slot. Progress Electromagn Res C 12:215–223
19. Deshmukh AA, Kumar G (2005) Compact broadband U-slot loaded rectangular microstrip antennas. Microwave Opt Technol Lett 46(6):556–559
20. Deshmukh AA, Kumar G (2005) Compact broadband E-shaped microstrip antennas. Electron Lett 41(18):989–990
21. Chair R, Mak CL, Lee KF, Luk KM, Kishk AA (2005) Miniaturized wide-band half U-slot and half E-shaped patch antennas. IEEE Trans Antennas Propag 53(8):2645–2652
22. Deshmukh A, Ray KP (2013) Analysis of broadband Ψ-shaped microstrip antennas. IEEE Antennas Propag Magaz 55(2):107–123
23. Deshmukh A, Ray KP (2015) Analysis of broadband variations of U-slot cut rectangular microstrip antennas. IEEE Antennas Propag Mag 57(2):181–193
24. Deshmukh A, Ray KP (2017) Analysis and design of broadband U-slot cut rectangular microstrip antennas. Sadhana—Academy Proceedings in Engineering Science, Springer Publication 42(10):1671–1684
25. Deshmukh A, Singh D (2018) On the design of wideband E-shape microstrip antennas on varying substrate thickness. In: Proceedings of ICWiCOM 2017, Mumbai, India
26. Deshmukh AA, Kumar G (2005) Compact broadband U-slot-loaded rectangular microstrip antennas 46(6):556–559. https://doi.org/10.1002/mop.21049
27. Kumar G, Ray KP (2003) Broadband microstrip antennas. Artech House, USA
28. Deshmukh AA, Chavali VAP (2023) Variations of square microstrip antenna loaded with U-slot and pair of rectangular slots for circular polarized response. Sādhanā 48:147. https://doi.org/10.1007/s12046-023-02190-4
29. IE3D simulation software, Version 12
30. CST Microwave Studio, Version 2019

Chapter 4
ANN Model of Rectangular and U-Slot Cut CMSA for Broadband Response

Abstract This chapter presents the application of artificial neural network models for the design of slot cut circular microstrip antennas aimed at achieving wideband performance. Unlike rectangular microstrip antennas, the field distribution in circular microstrip antennas is governed by Bessel functions, with the fundamental mode being TM_{11}. Although circular microstrip antennas inherently suppress harmonic resonances, their curved geometry limits direct gap coupling for bandwidth enhancement. Slot cut techniques offer an effective alternative, enabling wider bandwidth while maintaining compact antenna size. The chapter introduces artificial neural network models trained on simulation data over a broad frequency range (600–6000 MHz) and substrate thicknesses (0.02–0.1λ_g), using parameters such as patch radius, slot dimensions, and feed point location. A low-cost FR4 substrate with a suspended configuration is used throughout. The artificial neural network, implemented in Python, accurately predicts the optimal antenna dimensions for desired operating frequencies. Experimental validation confirms the effectiveness of the predicted designs, demonstrating broadband behaviour with minimal error. This approach provides a practical and efficient solution for designing advanced circular microstrip antennas using data-driven techniques.

4.1 Introduction

The CMSA is one of the regular-shaped MSA designs. In the case of RMSA, the field variation underneath the patch is defined in terms of sine or cosine functions. On the other hand, the fields for a CMSA are defined in terms of the Bessel function [1]. The fundamental mode of CMSA is TM_{11}. When compared with the RMSA, CMSA provides the harmonic suppression for the input signal supplied by the source, as resonance frequencies of CMSA are not related in the harmonic progression. As explained in Ref. [2], BW increment can be realized by using the gap coupling method. However, direct gap coupling of multiple CMSAs cannot realize wider BW as in RMSAs, as weak coupling between adjacent circular patches takes place due to their curvy geometry. Various stacked and gap-coupled designs of CMSA are

V. A. P. Chavali et al., *Neural Network Modelling of Microstrip Antennas*, Studies in Smart Technologies, https://doi.org/10.1007/978-981-96-9431-0_4

realized in Refs. [2, 3] for wider BW. However, the patch area, antenna volume, and substrate thickness are larger. Hence, slot cut technique can be used in CMSA to realize wideband or multi-band response, which also keeps the overall patch area smaller. The slot cut CMSA design in Ref. [4] was realized more than twice the BW when compared to the conventional CMSA. However, the specified slot shape is difficult to fabricate. Though various slot cut designs of CMSA are specified in the literature [5, 6], a very few proposed design methodologies of the antennas to predict the dimensions at a given frequency and substrate thickness using modern computation techniques [7]. Designing the effectual shape of a circular microstrip patch antenna for UHF band and C band is reported in Ref. [8]. Ansoft HFSS software is used for the simulation of antenna, and LM and Quasi-Newton BFG both are used to train the artificial neural network model. They are used to achieving the different resonating frequencies. But LM is faster and shows a little more accurate results than Quasi-Newton BFG. A 'Y' shape slotted structure loaded microstrip patch antenna is proposed in Ref. [9]. Here, RBF network which is a feed-forward neural network is used. To design and analyse the proposed antenna, HFSS software tool has been utilized. The analysed design exhibits multi-band behaviour at different resonant frequencies. However, the size of the data sets can be increased for better accuracy. Also, the computational time can be improved for the proposed model. In the circular radiating patch with a ring slot at the centre discussed in Ref. [10], three different ML algorithms are applied to the collected data for prediction of resonance frequency.

This chapter discusses the ANN models of slot cut broadband designs of CMSA. In case of suspended dielectric substrate, low-cost lossy FR4 substrate ($\varepsilon_r = 4.3$, $h = 0.16$ cm, $\tan\delta = 0.02$) located above the ground plane using finite air gap (h_a) has been used. Initially, the functioning of both antennas with respect to various resonant modes leading to the broadband response is explained followed by the explanation of its ANN model in detail. In the ANN modelling for each configuration, ANN is trained using various data sets of large frequency range and substrate thickness. These data sets include various parameters of antenna such as thickness of the substrate, resonance frequency as function of operating wavelength, total thickness of the substrate, patch dimensions such as radius, slot length, slot width, slot position, and feed point location obtained using simulation for the given resonance frequency, and the half wavelength value at the considered frequency. The data sets for training ANN are included at the frequency intervals of 200 MHz over 600–6000 MHz frequency band. The ANN code is written using programming language 'Python' [11]. Next utilizing trained ANN model, dimensions of U-slot cut CMSA are predicted over thickness of the substrate in the range of 0.02–0.1λ_g and across 600–6000 MHz frequency spectrum. An experimental validation for predicted results is carried out which shows a wideband response with a greater degree of accuracy. Antenna dimensions in the tables throughout this chapter are specified in cm, and frequencies in MHz. All proposed antennas are simulated using CST [12], and IE3D [13].

4.2 ANN Model of Rectangular and U-Slot Cut CMSA for Broadband Response

4.2.1 Rectangular Slot Cut CMSA for Broadband Response

The design of RMSA loaded with a pair of rectangular slots realizes wider BW due to an optimum inter-spacing between TM_{10} and modified TM_{02} resonant modes. A similar design of rectangular slot cut CMSA is reported in Ref. [14]. The slots cut CMSA realizes measured BW of around 194 MHz (22.3%) with a peak gain of 9 dBi [14]. However, the resonant modes responsible for the BW enhancement were not identified and explained. The polygon shaped MSAs with rectangular slots in Ref. [15] identified the first and second order modes are responsible for the wider BW in the polygon shape MSAs, ranging from triangular to circular geometry. It was put forward in that study that resonant modes responsible for the BW enhancement in rectangular slot cut CMSA are TM_{11} and TM_{21}. As specified in Ref. [15], with the length increment in the rectangular slots, the TM_{21} mode frequency decreases as its modal surface currents are orthogonal to the rectangular slot length. Thus, with space and impedance tuning between TM_{11} and TM_{21} resonant modes, BW enhancement takes place.

To study the BW response in the pair of slots cut CMSA, it is designed at 900 MHz on air substrate for '$h_a = 0.05\lambda$', as shown in Fig. 4.1a, b. The rectangular slot dimensions such as l_s, w_s, y, and x_f are optimized to realize wideband response. Antenna dimensions in the optimum design are $R = 8$, $l_s = 0.9$, $w_s = 1$, $y = 2.7$, and $x_f = 5.4$ cm. This design yields simulated reflection coefficient (S_{11}) BW of 185 MHz (20.8%), with a peak gain of 8.9 dBi, as shown in Fig. 4.1c. As the configuration is coaxially fed, a single loop due to the coupling of TM_{11} and TM_{21} resonant modes is observed inside VSWR = 2 circle in the Smith chart.

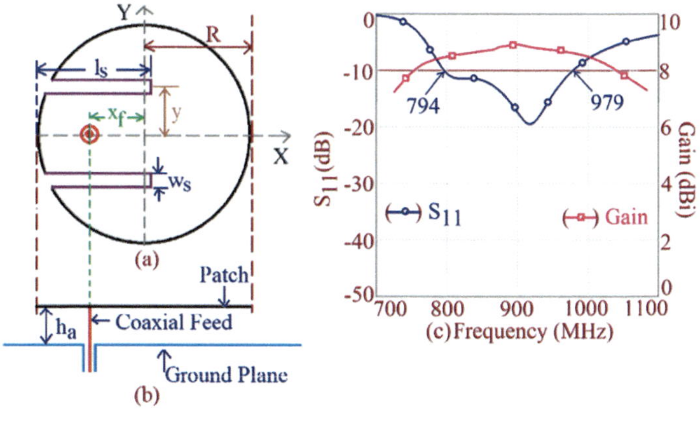

Fig. 4.1 **a**, **b** Rectangular slot-loaded CMSA, simulated **c** Smith chart, and **d** S_{11}, gain plots

With the above understanding of the geometry, ANN modelling for a pair of rectangular slot cut CMSA is presented. To design the ANN model, the crucial stage is training the ANN through the data sets. For this initially using air substrate, the CMSA with a pair of rectangular slots is optimized for broadband response by tuning the spacing between TM_{11} and TM_{21} modes, for each design over a frequency range of 900–5500 MHz, and in frequency intervals of 200 MHz. At each frequency, CMSA with rectangular slots is optimized for substrate heights, varying in the range 0.05λ–0.08λ. The antenna design is carried out in the following manner for each frequency. For each substrate thickness, to create training data set for the ANN, the detail method of data set creation is represented in the flowchart as shown in Fig. 4.2.

1. Design CMSA for a particular TM_{11} mode frequency f_r, and substrate thickness h using resonance frequency equation of CMSA [1].
2. Feed the CMSA at appropriate location starting at (R/2,0).
3. Optimize the radius of the CMSA to tune the resonant peak to f_r.

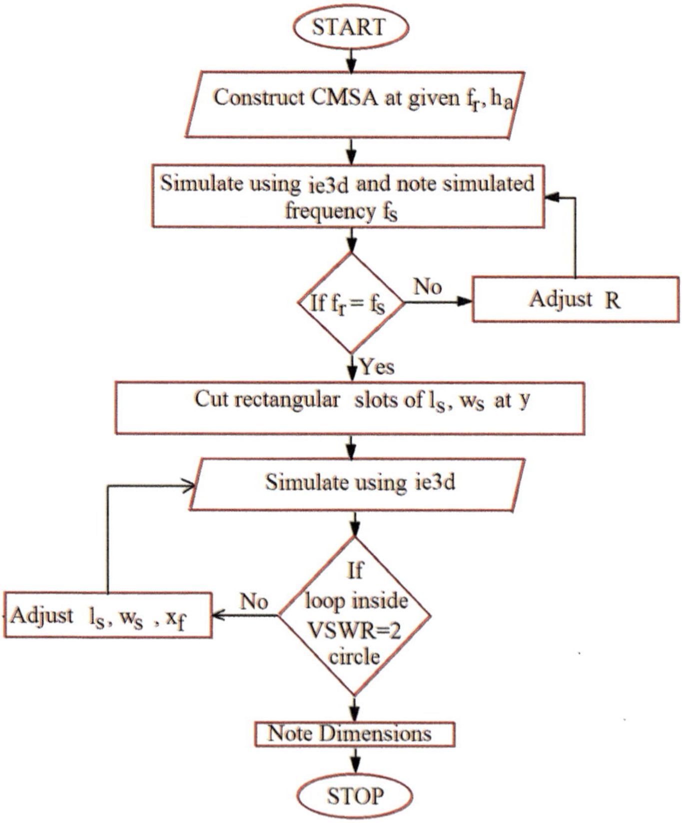

Fig. 4.2 Flowchart of rectangular slot cut CMSA design

Table 4.1 Sample training data set for rectangular slot cut CMSA

f_r	$\lambda/2$	h_a	R	x_f	l_s	w_s	Y	f_1	f_2	BW	$\Delta R/2$
600	25	3.5	11.5	– 6.8	12.2	1	3.3	599.9	696.1	96.2	0.28
700	21.4	2.10	10.5	– 3.9	8	0.9	3.0	724.8	802.7	77.9	0.10
900	16.7	1.67	8.15	– 3.2	6.2	0.8	2.4	928	1066.8	138.8	0.11
1500	10	1.00	4.88	– 1.5	4	0.6	1.4	1532	1726.5	194.5	0.12
2100	7.14	1.00	3.34	– 1.3	2.7	0.5	0.95	2221.7	2470.9	249.2	0.23
2700	5.55	0.56	2.71	– 0.9	2.1	0.4	0.8	2759	3143.7	384.7	0.12
3300	4.54	0.64	2.12	– 0.7	1.7	0.3	0.65	3410	3899.6	489.6	0.24
3900	3.84	0.54	1.79	– 0.5	1.46	0.2	0.5	4019	4510.7	491.7	0.24
4300	3.48	0.35	1.69	– 0.5	1.3	0.2	0.5	4385	5034.1	649.1	0.15
4900	3.06	0.43	1.4	– 0.4	1.13	0.2	0.4	5115	5823	708	0.30
5100	2.94	0.30	1.4	– 0.4	1.04	0.4	0.3	5467.2	6351.6	884.4	0.23
5500	2.73	0.38	1.25	– 0.4	1.6	0.3	0.25	5640.7	6156.4	515.7	0.29

4. Further, optimize the slot dimensions l_s, w_s so that the loop due to the coupling of TM_{11} and TM_{21} modes lies inside the VSWR = 2 circle.

The sample training data set of single-stage ANN model created for the frequency range of 900–5500 MHz on varying substrate thickness forms 0.05λ–0.08λ is shown in Table. 4.1. The negative values of x_f in the table represent the feed point location on the left side of the centre of the CMSA.

The ANN model for rectangular slot cut CMSA is designed using feed-forward supervised networks with a backpropagation algorithm in two different designs, i.e., single-stage ANN model and double-stage ANN model, which are explained below. In both the models, frequency range is considered 900–5500 MHz, and substrate thicknesses are considered in the range 0.05λ–0.08λ.

4.2.1.1 Single-Stage ANN Model

After training the network, it was tested on different frequencies for different substrate thicknesses. The neural network tool (nntool) in MATLAB helps to select data, create and train a network, and give output and error values. Neural Fitting tool (nftool) helps in curve fitting of the input and target values and analyse the performance of the neural network using mean squared error and regression analysis. The average of square of differences between output values and target values is called MSE. The lower the values, the better the result. If the result is zero, then there is no error. The measurement of extent to which outputs and targets are correlated is called regression values. If the value is 1 then the outputs and targets have a close relationship. If the value is closer to 0 then they have a random relationship. The block diagram of the single-stage ANN model and the nodal structure of single-stage ANN model of pair rectangular slot cut CMSA are shown in Fig. 4.3a, b.

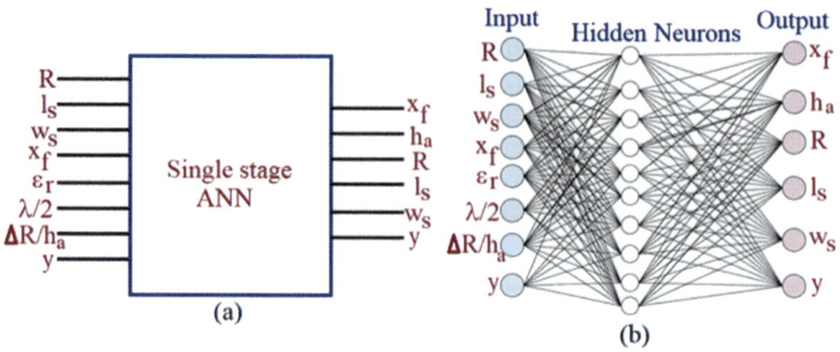

Fig. 4.3 **a** Block diagram, **b** nodal structure of single-stage ANN model of rectangular slot cut CMSA

The input parameters applied to the neural network are the resonant frequency f_r of CMSA, and its corresponding half wavelength, dielectric constant ε_r, slot dimensions l_s, w_s, feed point location x_f, and the ratio $\Delta R/h_a$. To determine the number of hidden neurons and hidden layers, a trial-and-error method is used. A specific number of hidden neurons and layers are selected for the network, and the model is then trained on the data samples. If the value of MSE obtained is low, it is noted down along with the model parameters, whereas, if the value of MSE is large, the number of neurons and layers are changed until the lowest value of MSE is obtained. The estimated outputs of this network are x_f, h_a, R, l_s, w_s, and slot position from the centre axis, y. The details of layers, number of nodes per layer, and the algorithm used for training are shown Table. 4.2. The plot between the MSE and number of iterations is indicated as performance plot. The best results for validation, testing, and training are plotted in this. Mean squared error is the average squared difference between the targets and outputs. The lower the value of MSE, more is the effectiveness of the network. The performance plot for the model is shown in Fig. 4.4. It can be observed that the best validation performance is 0.41242 at Epoch 29. The dimensions predicted by the single-stage ANN architecture are given in Table. 4.3, after minor changes. These minor changes in the dimensions are done to obtain practically realizable values.

Further, from the dimensions observed from the table, the rectangular slot-loaded CMSA is designed, and simulations are conducted to verify the result in terms of S_{11} BW. The results for the design at 1500 MHz are shown in Fig. 4.5.

Table 4.2 Details of ANN of rectangular slot cut CMSA

S. No.	Parameter	Value
1	Number of inputs	8
2	Number of outputs	6
3	Number of hidden layers	1
4	Number of hidden neurons	10
5	Number of epochs	1000
6	Training algorithm	LM

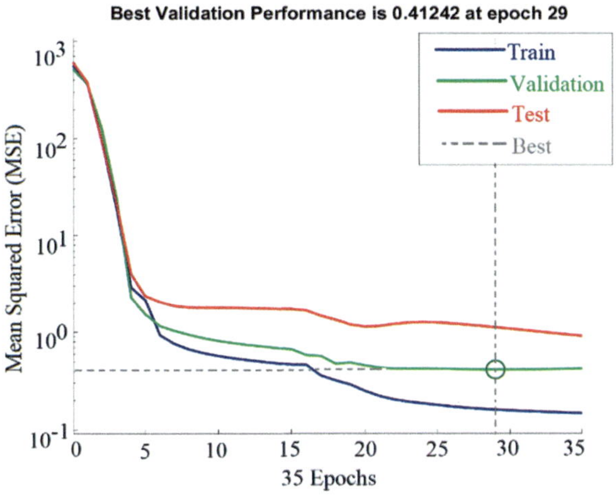

Fig. 4.4 Performance plot of single-stage ANN model of rectangular slot cut CMSA

A pair of rectangular slot cut CMSA designed on air substrate at 1500 MHz at $h = 0.05\lambda$ yields simulated and measured S_{11} BW of 173 MHz (10.49%) and 184 MHz (11.16%), respectively with a peak gain of 9.5 dBi. Antenna fabricated prototype is shown in Fig. 4.6.

The simulation of pair of rectangular slot-loaded CMSA, whose dimensions are selected from Table 4.3, shows that the loop size achieved in the VSWR = 2 circle was smaller in size, indicating a lower coupling between the resonant modes of slot cut patch thus giving a narrower S_{11} BW. To enhance BW, on the neural network outputs, significant parametric optimization is needed further. The demand on the neural network to ascertain relationships between input parameters is further increased by the huge number of input parameters in single-stage ANN. As a result, it had an impact on neural network performance, requiring more time and processing resources to anticipate output parameter values precisely enough to provide a correct loop size. Thus, to overcome this drawback, two-stage ANN is proposed.

4.2.1.2 Two-Stage ANN Model

A two-stage ANN is a combination of two different ANN models. It is an improved version of the single-stage ANN model. The first model in this method uses input frequency (f_r), half-wavelength ($\lambda/2$), dielectric constant (ε_r) of substrate, substrate thickness (h_a), and $\Delta R/h_a$ value as input parameters. Radius (R) of the circular patch is the estimated output of the network. This optimized value of radius is then used as an input parameter in the second model of the process, along with slot length (l_s), slot width (w_s), slot separation (y), feed position (x_f), dielectric constant (ε_r), half-wavelength ($\lambda/2$), and the $\Delta R/h_a$ value. The estimated outputs of the second model are x_f, h_a, l_s, w_s, y, and R. Both the models used LM algorithm for training the data. Both the models consist of 10 hidden neurons and were trained on 1000 epochs. The

Table 4.3 Predicted data set for rectangular slot cut CMSA with single-stage ANN

f_r	h_a/λ	x_f	h_a	l_s	w_s	Y	R
600	0.05	– 3.5	2.6	4.5	0.8	6.8	9
600	0.07	– 4.5	2.9	10.2	0.8	6.5	9
700	0.05	– 3.4	2	5.1	0.8	6.8	9
700	0.07	– 4.5	2.8	10.3	0.8	5.5	8.9
900	0.05	– 3.2	1.5	6.2	0.75	5.05	8.1
900	0.07	– 2.9	2.2	7.5	0.7	4.9	8
1500	0.05	– 1.6	1	4	0.6	2.8	4.8
1500	0.07	– 1.4	1.4	3.8	0.6	2.6	4.6
2100	0.05	– 1.3	0.75	2.6	0.5	1.9	3.5
2100	0.07	– 1.3	0.9	2.6	0.5	1.9	3.4
2700	0.05	– 0.9	0.6	1.95	0.4	1.6	2.7
2700	0.07	– 0.9	0.7	2	0.4	1.4	2.6
3300	0.05	– 0.7	0.5	1.65	0.3	1.3	2.2
3300	0.07	– 0.7	0.6	1.7	0.2	1.3	2.1
3900	0.05	– 0.6	0.5	1.5	0.2	1.2	1.85
3900	0.07	– 0.5	0.5	1.4	0.2	1	1.8
4300	0.05	– 0.5	0.4	1.3	0.2	1	1.65
4300	0.07	– 0.5	0.5	1.3	0.2	0.9	1.6
4900	0.05	– 0.5	0.4	1.2	0.2	0.9	1.5
4900	0.07	– 0.5	0.4	1.2	0.2	0.9	1.45
5100	0.05	– 0.5	0.4	1.2	0.3	0.8	1.5
5100	0.07	– 0.6	0.45	1.55	0.2	0.8	1.5
5500	0.05	– 0.5	0.4	1.2	0.3	0.8	1.45
5500	0.07	– 0.6	0.45	1.4	0.2	0.8	1.4

block diagram and nodal structure of two-stage ANN model are shown in Fig. 4.7a, b.

The parameters for both the models are shown in Table 4.4. The model accuracy can be increased by increasing the number of hidden neurons. However, using too many hidden neurons may lead to an overfitting of the model. Performance plots for both the models are shown in Fig. 4.8a, b. For the first model, the best validation performance is 0.015928 at epoch 172 and for the second model, it is 0.51115 at epoch 70.

For training of the double-stage ANN model, the same data set which is used for single-stage ANN model is considered which is given in Table. 4.1. The dimensions predicted by the two-stage ANN architecture are given in Table. 4.5, after minor changes. These minor changes in the dimensions are done to obtain practically realizable values.

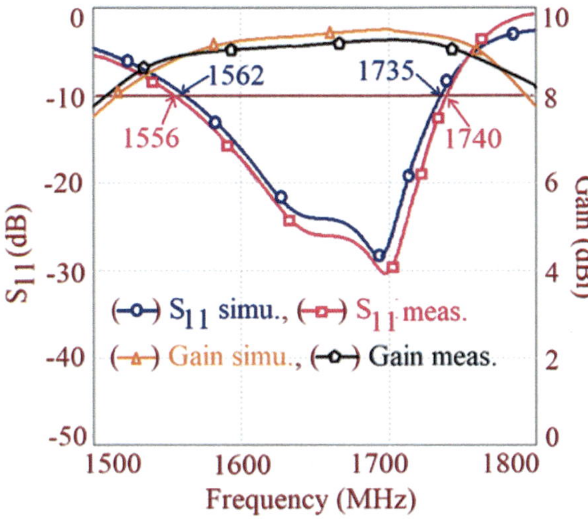

Fig. 4.5 S_{11} BW and gain plots for pair of rectangular slot cut CMSA designed on air substrate at 1500 MHz using single-stage ANN

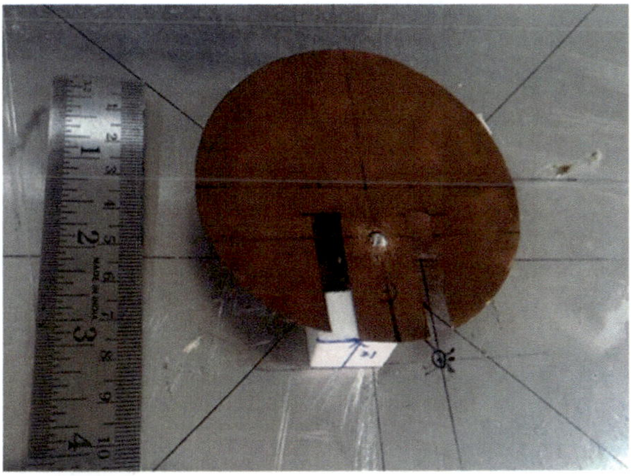

Fig. 4.6 Fabricated prototype of pair of rectangular slot cut CMSA on air-suspended substrate at 1500 MHz using single-stage ANN

Further, from the dimensions observed from the above table, the pair of rectangular slot-loaded CMSA is designed, and simulations and measurements are conducted to verify the result in terms of S_{11} BW. The results for the same are shown in Fig. 4.9a. For the pair of rectangular slot-loaded CMSA designed on air substrate for TM_{11} mode frequency of 1500 MHz at $h_a = 0.07\lambda$, simulated and measured BW for $S_{11} \leq -10$ dB is 272 MHz (16.7%), 288 MHz (17.7%), respectively. The antenna offers peak broadside gain of about 8.8 dBi. Antenna fabricated prototype on air-suspended substrate is shown in Fig. 4.9b.

To summarize, ANN modelling for a pair of rectangular slot cut CMSA using two methods, viz. single-stage, and two-stage ANN is proposed. A comparison of simulated impedance BW of rectangular slot cut CMSA designed with predicted dimensions of single-stage and double-stage ANN models is given in Table. 4.6.

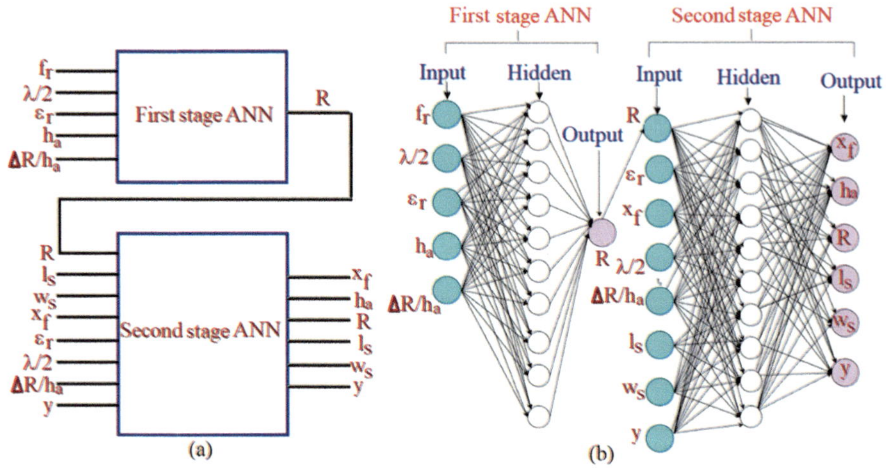

Fig. 4.7 **a** Block diagram, **b** nodal structure of two-stage ANN model of rectangular slot-loaded CMSA

Table 4.4 Model parameters in both stages of two-stage ANN model

Parameter	Values in first stage	Values in second stage
Number of inputs	5	8
Number of outputs	1	6
Number of hidden layers	1	10
Number of hidden neurons	10	1
Number of epochs	1000	1000
Training algorithm	LM	LM

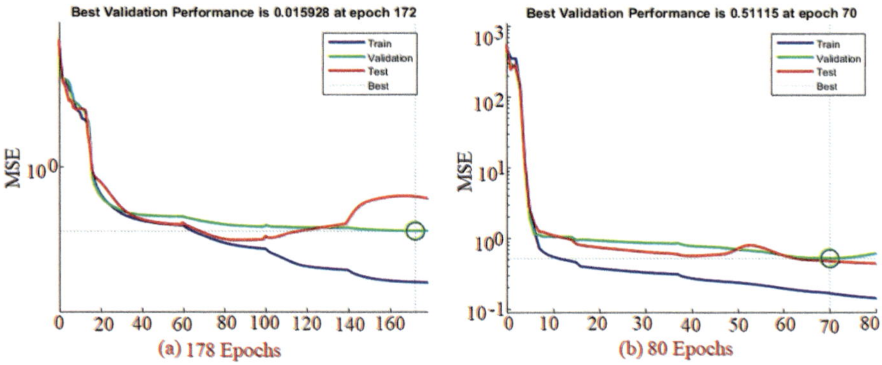

Fig. 4.8 Performance of ANN model: **a** first stage, **b** second stage

Table 4.5 Two-stage ANN optimized practical dimensions

f_r	h_a/λ	x_f	h_a	l_s	w_s	Y	R
900	0.05	– 2.9	1.7	6.2	0.8	5.1	8.4
1500	0.05	– 1.6	1.1	3.7	0.6	2.8	5.0
1500	0.07	– 1.5	1.4	3.7	0.6	2.6	4.6
2100	0.05	– 1.4	0.8	2.7	0.5	2.0	3.6
2700	0.05	– 1.1	0.6	2.2	0.4	1.7	2.8
3300	0.05	– 0.8	0.5	1.7	0.3	1.3	2.2
3300	0.07	– 0.8	0.6	1.7	0.3	1.3	2.1
3900	0.05	– 0.6	0.4	1.4	0.2	1.2	1.8
5100	0.05	– 0.5	0.3	1.1	0.2	0.8	1.5
5100	0.07	– 0.4	0.4	1.15	0.2	0.8	1.5
5500	0.05	– 0.5	0.3	1.1	0.2	0.8	1.4
5500	0.07	– 0.4	0.4	1.1	0.2	0.8	1.4

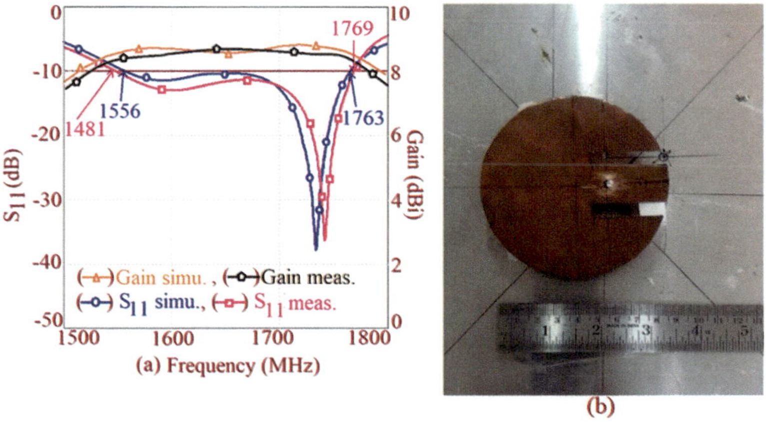

Fig. 4.9 **a** S_{11} BW and gain plots, and **b** fabricated prototype of pair of rectangular slot-loaded CMSA designed on air substrate at 1500 MHz using two-stage ANN

Comparatively wider BW was observed in two-stage ANN model, due to better coupling of the resonant modes. Fewer adjustments are required in the predicted ANN parameters as values obtained are not practically realizable for many antenna parameters. Neural network gave better accuracy in two-stage ANN as responsibility of prediction was divided into two stages. Thus, at the same time and computational power used, two-stage ANN performed better giving it an advantage in scalability to include more data points and a wider frequency range compared to single-stage ANN.

Table 4.6 Comparison of single-stage and two-stage ANN models

f_r	h_a	Impedance BW (in MHz)	
		Single-stage ANN	Two-stage ANN
900	0.05λ	0.0855	0.1345
900	0.07λ	Impedance loop outside VSWR = 2 circle	Impedance loop outside VSWR = 2 circle
1500	0.05λ	0.1937	0.247
1500	0.07λ	0.203	0.1919
2100	0.05λ	0.3175	0.2898
2100	0.07λ	0.2676	0.2629
2700	0.05λ	0.4413	0.3943
2700	0.07λ	0.4052	0.3599
3300	0.05λ	0.5068	0.546
3300	0.07λ	0.4269	0.5038
3900	0.05λ	0.545	0.648
3900	0.07λ	0.523	0.476
4300	0.05λ	0.637	0.706
4300	0.07λ	0.54	0.628
4900	0.05λ	0.62	0.732
4900	0.07λ	0.684	0.728
5100	0.05λ	0.563	0.836
5100	0.07λ	Impedance loop outside VSWR = 2 circle	Impedance loop outside VSWR = 2 circle
5500	0.05λ	0.825	Impedance loop outside VSWR = 2 circle
5500	0.07λ	Impedance loop outside VSWR = 2 circle	Impedance loop outside VSWR = 2 circle

4.2.2 U-Slot Loaded CMSA for Broadband Response

Design of CMSA with U-slot with superstrate which acts as protective cover to the patch is reported in Ref. [16]. However, the functioning of the antenna with respect to resonant modes is not explained. In this section broadband behaviour of CMSA with U-slot is explained through parametric variation in the U-slot dimensions and variation in corresponding resonant modes.

The configuration of CMSA loaded with U-slot is shown in Fig. 4.10a, b. Initially, a CMSA for its fundamental TM_{11} mode frequency of 900 MHz on air substrate with $h_a = 0.07\lambda$ is designed, and its radius is calculated using CMSA resonance frequency equation [1]. To obtained the broadband response, the horizontal length L_h and vertical slot length L_v of the U-slot are parametrically optimized. The variations in L_h and L_v tunes TM_{21} mode frequency with respect to TM_{11} mode as shown in Fig. 4.10c, d that achieves wider BW.

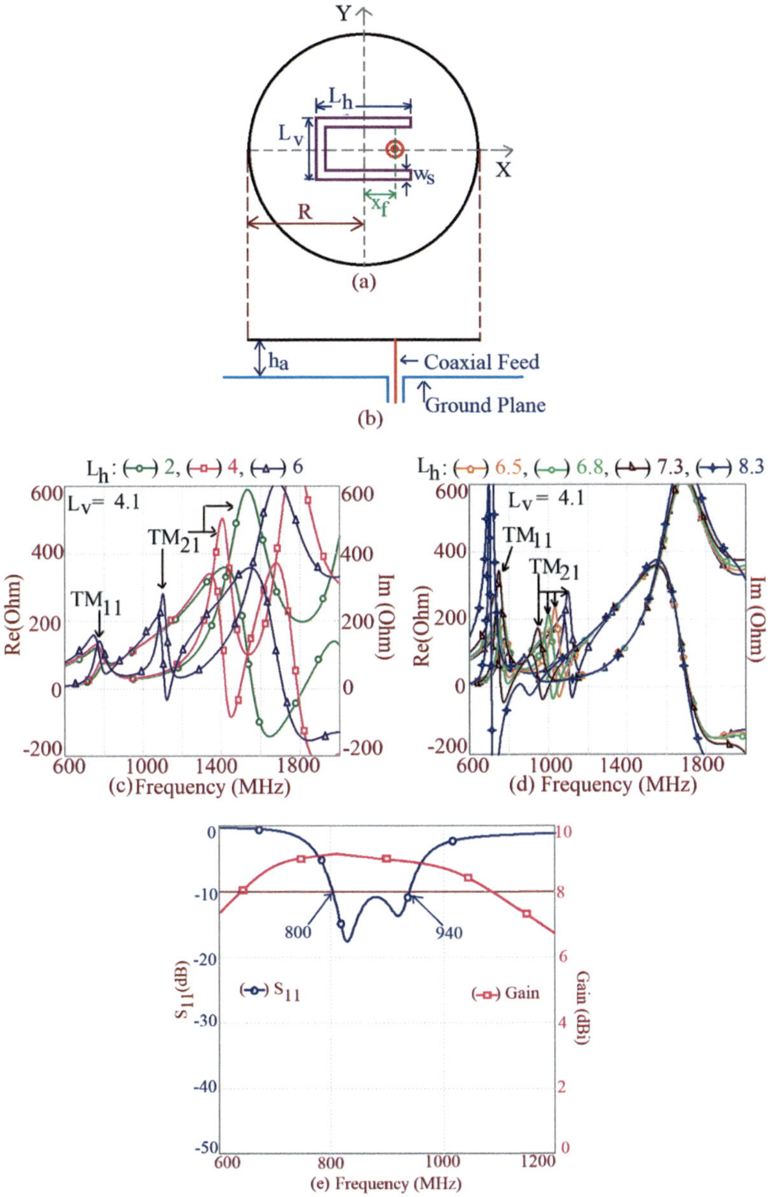

Fig. 4.10 **a, b** CMSA with U-slot, its simulated **c, d** resonance frequency plots for varying slot length in U-slot, and **e** S_{11} BW and gain plots for wideband design of CMSA with U-slot

The dimensions in the optimum configuration are $R = 8.8$, $L_v = 4.1$, $L_h = 8.3$, $w_s = 1$, $h_a = 2.4$, and $x_f = 2.5$ cm. The simulated results of the CMSA with U-slot optimized at $f_{TM11} = 900$ MHz are shown in Fig. 4.10e. The simulated BW is 140 MHz (16%) with a peak broadside gain of 9.13 dBi is observed.

Further, to design the ANN model, training data set is generated. For this initially, the CMSA with U-slot is optimized for broadband response by tuning the space between TM_{11} and TM_{21} modes over a frequency range of 800 MHz–5100 MHz in frequency intervals of 600 MHz. Here air is used as the substrate. For each frequency, CMSA with U-slot is optimized for substrate heights, varying in the range 0.05λ–0.08λ. The antenna design is carried out in the following manner for each frequency and for each substrate thickness to create training data set for the ANN model, and the detailed method of data set creation is represented in the flowchart shown in Fig. 4.2.

1. Design CMSA for a particular TM_{11} mode frequency f_r, and substrate thickness h using resonant frequency equation of CMSA [2].
2. Feed the CMSA at appropriate location starting at (R/2,0).
3. Optimize the radius of the CMSA to tune the resonant peak to f_r.
4. Further, optimize the slot dimensions L_v, L_h, w_s, and x_f so that the loop due to the coupling of TM_{11} and TM_{21} modes lie inside the VSWR = 2 circle.

The sample training data set of single-stage ANN model created for the frequency range of 800 MHz–5100 MHz on varying substrate thickness forms 0.05λ–0.08λ is shown in Table. 4.7.

Further, the training data set created using the above method is used to train the ANN model. As single-stage ANN model results in some drawbacks, as mentioned in the previous discussion, here dimensions of CMSA with U-slot are predicted using two-stage ANN model. The first model in this method uses input frequency (f_r), half-wavelength ($\lambda/2$), dielectric constant (ε_r) of substrate, substrate thickness (h_a), and $\Delta R/h_a$ value as input parameters. Radius (R) of the circular patch is the estimated output of the network. This optimized value of radius is then used as an input parameter in the second model of the process, along with U-slot horizontal slot length (L_h), vertical slot length (L_v), slot width (w_s), substrate thickness (h_a), feed position (x_f), dielectric constant (ε_r), half-wavelength ($\lambda/2$), and the $\Delta R/h_a$ value. The estimated outputs of the second model are L_v, L_h, x_f, h_a, w_s, and R. Both the models used Levenberg–Marquardt (LM) algorithm for training the data. Both the models consist of 10 hidden neurons and were trained on 1000 epochs. The block diagram and nodal structure of two-stage ANN models are like the ones of CMSA with a pair of rectangular slots shown in Fig. 4.7a, b, with corresponding slot parameter modifications. The parameters for both the models are shown in Table 4.4. The model accuracy can be increased by increasing the number of hidden neurons, but using too many hidden neurons may lead to overfitting of the model. Performance plots for both the models are shown in Fig. 4.11a, b. For the first model, the best validation performance is 0.19667 at epoch 143 and for the second model, it is 0.12387 at epoch 28. After training the ANN with the training data set, the ANN is used to predict the dimensions of U-slot CMSA by feeding the ANN with new parameters.

Table 4.7 Sample training data set of CMSA with U-slot

f	h_a	R	x_f	L_h	w_s	L_v	BW	$\Delta R/h_a$
800	1.88	8.99	2.5	8	1	4.2	170	0.21
1000	1.8	7.2	1.7	6.7	1	4	167	0.17
1200	1.75	5.85	1.1	6	1	3.2	173	0.23
1400	1.5	5.02	0.9	5.1	0.8	2.8	192	0.22
1600	0.94	4.55	1.3	4.3	0.8	2.4	263	0.15
1800	1.17	3.86	1	3.5	0.5	2	333	0.26
2000	0.9	3.55	0.9	3.2	0.4	1.8	436	0.22
2200	0.68	3.4	0.4	3.2	0.4	1.8	329	0.01
2400	0.75	3	0.4	2.9	0.4	1.8	386	0.17
2600	0.69	2.7	0.7	2.6	0.4	1.4	421	0.27
2800	0.64	2.5	0.6	2.4	0.4	1.4	490	0.28
3000	0.6	2.34	0.5	2.2	0.3	1.5	700	0.27
3200	0.47	2.28	0.8	2	0.3	1.1	451	0.14
3400	0.53	2.1	0.5	2	0.3	1.3	759	0.20
3600	0.42	2.02	0.7	1.8	0.3	1.1	542	0.15
3800	0.55	1.82	0.2	1.8	0.2	1.2	1101	0.28
4000	0.38	1.82	0.5	1.6	0.2	1	551	0.15
4200	0.5	1.6	0.6	1.4	0.2	0.8	720	0.37
4400	0.34	1.68	0.6	1.4	0.2	0.8	654	0.07
4600	0.33	1.55	0.5	1.4	0.2	0.8	575	0.25
4800	0.31	1.51	0.5	1.4	0.2	0.8	425	0.17
5000	0.42	1.36	0.4	1.4	0.2	0.8	580	0.33

The dimensions predicted by the two-stage ANN architecture are given in Table 3. 9, after minor changes. These minor changes in the dimensions are done to obtain practically realizable values. In Table 4.8 all frequencies are specified in MHz, and all dimensions are mentioned in cm.

Further, from the dimensions observed from the above table the U-slot loaded CMSA is designed, and simulations are conducted to verify the result in terms of S_{11} BW plot. The results are shown in Fig. 4.12a. The simulated and measured S_{11} BW of 206 MHz (12.15%), 234 MHz (13.7%) are realized respectively with a peak gain of about 9.04 dBi by the U-slot loaded CMSA designed on air substrate at 1500 MHz at $h = 0.06\lambda$. The fabricated prototype of the U-slot CMSA is shown in Fig. 4.12b.

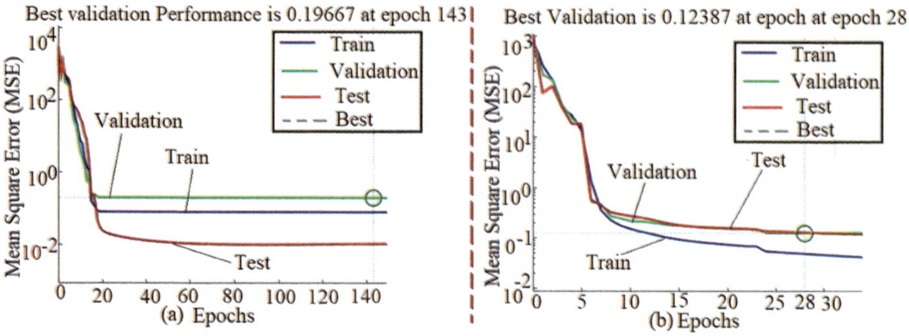

Fig. 4.11 Performance of two-stage ANN model: **a** first stage, **b** second stage

Table 4.8 Sample set of two-stage ANN optimized practical dimensions for U-slot CMSA

f	h_a	h_a	x_f	L_h	w_s	L_v	R
600	0.05λ	2.6	2.5	8.6	1.0	4.2	9.0
600	0.06λ	2.6	2.5	8.6	1.0	4.2	9.0
600	0.07λ	2.6	2.5	8.6	1.0	4.2	9.0
700	0.05λ	2.3	2.7	8.2	1.0	4.2	9.0
700	0.06λ	2.2	2.7	8.2	1.0	4.2	9.0
700	0.07λ	2.2	2.7	8.2	1.0	4.2	9.0
900	0.07λ	2.4	2.5	8.3	1.0	4.1	8.8
2100	0.05λ	1.1	1.3	3.2	0.5	1.5	3.4
2700	0.06λ	0.8	0.8	2.5	0.4	1.4	2.6
3300	0.05λ	0.4	0.3	2.7	0.2	1.4	2.7
4900	0.05λ	0.3	0.5	1.6	0.2	0.8	1.8
5500	0.06λ	0.4	0.4	1.4	0.2	0.8	1.4
5500	0.07λ	0.4	0.4	1.4	0.2	0.8	1.4

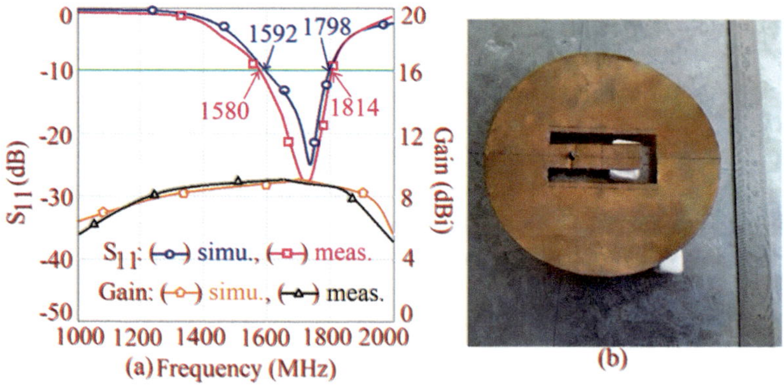

Fig. 4.12 **a** S_{11} BW and gain plots, and **b** fabricated prototype of U-slot loaded CMSA designed on air substrate at 1500 MHz using two-stage ANN

4.3 Summary

ANN models for slot cut CMSA are presented in this chapter. Initially two techniques of ANN modelling were proposed for rectangular slot-loaded CMSA. However, the slot cut CMSA designed using single-stage ANN predicted dimensions realized narrow BW due to low coupling between the resonant modes, exhibiting a smaller loop size. Thus, two-stage ANN model was proposed, which enhanced BW due to the increased number of parameters in each stage of ANN network. Further, two-stage ANN model is continued in the design of U-slot loaded CMSA. Both the techniques of ANN utilized LM algorithm for reduced proceeding time and accurate results. In all the MSAs, ANN predicted dimensions and calculated dimensions have shown less than 5% variation.

References

1. Kumar G, Ray KP (2003) Broadband microstrip antennas. Artech House, Norwood, MA, USA
2. Datta M, Kumar G (2000) Broadband gap-coupled circular microstrip antennas. In: IEEE antennas and propagation society international symposium. Transmitting waves of progress to the next millennium. 2000 digest. Held in conjunction with: USNC/URSI national radio science meeting, vol 3. IEEE, pp 1418–1421
3. Maurya RK et al (2019) ANN and ANFIS model for design of triple band, DGS triangular ring micro-strip antenna. IEEE, New Delhi
4. Chavali VAP, Deshmukh AA (2021) Modified designs of U-slot cut microstrip antennas for wider bandwidth. IETE J Res 69(6):3540–3549. https://doi.org/10.1080/03772063.2021.1914203
5. Singh AK, Gangwar RK, Kanaujia BK (2015) Orthogonal slot-loaded coaxially stacked annular ring antenna with circular patch for multiband application. J Electromagn Waves Appl 29(12):1630–1643.https://doi.org/10.1080/09205071.2015.1054956
6. Kulkarni SS, Mishra A (2023) Circular microstrip antenna inspired by metamaterial and U-shaped notch for ISM applications. In: IEEE wireless antenna and microwave symposium (WAMS), Ahmedabad, India, pp 1–5.https://doi.org/10.1109/WAMS57261.2023.10242873
7. Deshmukh AA, Shukla M, Patel S, Labde S, Chavali VAP (2020) Artificial neural network model for calculating the dimensions of circular and semi-circular microstrip antennas for given resonance frequency. In: Vasudevan H, Gajic Z, Deshmukh A (eds) Proceedings of international conference on wireless communication. Lecture notes on data engineering and communications technologies, vol 36. Springer, Singapore. https://doi.org/10.1007/978-981-15-1002-1_22
8. Singh G, Mittal R (2017) Artificial neural network based analysis of circular multiband antenna. IJLTEMAS
9. Singh JP et al (2014) Design analysis of a hexaband slot loaded microstrip patch antenna using ANN. Int J Comput Appl
10. Sharma K, Pandey GP (2019) Predicting resonant frequency of annular ring compact microstrip antenna using various machine learning techniques. IEEE, Rajkot
11. http://www.python.org/
12. CST Microwave Studio, Version 2019
13. IE3D simulation software, Version 12
14. Deshmukh AA, Kumar G (2006) Various slot loaded broadband and compact circular microstrip antennas. Microw Opt Technol Lett 48(3):435–439

15. Deshmukh AA, Chavali VAP (2021) Wideband pentagonal shape microstrip antenna using a pair of rectangular slots. Prog Electromagn Res 107:113–126. https://doi.org/10.2528/pierc20102602
16. Luk KM, Lee KF, Tam WL (1997) Circular U-slot patch with dielectric superstrate. Electron Lett 33:1001–1002

Chapter 5
ANN Modelling of Compact Slot Cut MSAs for Broadband and CP Response

Abstract Microstrip antennas are widely valued for their low profile, lightweight structure, ease of fabrication, and adaptability to mounting surfaces. However, their inherently narrow bandwidth and the demand for compact designs in modern mobile communication systems pose significant design challenges. While techniques such as slot cutting, shorting posts, and the use of high-permittivity substrates enable antenna miniaturization, they often lead to reduced bandwidth. Conversely, bandwidth enhancement strategies like stacked or gap-coupled multi-resonator designs increase antenna size and complexity. A promising alternative is the incorporation of resonant slots within the patch, which preserves the low-profile form while bringing higher order modes closer to the fundamental mode, thus enhancing bandwidth. For applications in wireless and personal communication, antennas with compact size, broad bandwidth, and circular or elliptical polarization are particularly advantageous, mitigating signal degradation from multi-path effects. This chapter presents Artificial Neural Network models for a range of compact slot cut microstrip antenna configurations including rectangular microstrip antennas, square microstrip antennas, and semi-circular microstrip antennas targeted at achieving wideband and circularly polarized responses. Each antenna's operation is first analysed in terms of its resonant and orthogonal modes, followed by detailed artificial neural network-based design modelling.

5.1 Introduction

The appealing qualities of the conventional MSA are low profile, light weight, simple manufacturing, and conformability to the mounting hosts [1]. Typically, a conducting patch is printed on a grounded microwave substrate. Yet, because MSAs are intrinsically narrow bandwidth (BW) elements, practical applications typically require BW augmentation. Furthermore, lower antenna sizes are typically needed for applications in today's mobile communication systems to satisfy the miniaturization demands of the mobile devices. Therefore, for a real-world MSA applications, size reduction and BW increase are increasingly important design features. One can achieve

V. A. P. Chavali et al., *Neural Network Modelling of Microstrip Antennas*, Studies in Smart Technologies, https://doi.org/10.1007/978-981-96-9431-0_5

a small-sized MSA by cutting a slot, utilizing a shorting post or a substrate with a higher dielectric constant [1–3]. The compact MSA realized using any of these techniques possesses narrow BW. Either a thicker substrate with lower dielectric constant or multi-resonator gap-coupled and stacked designs enhance the bandwidth (BW) in compact designs [2]. These BW enhancement methods enlarge or thicken the antenna. The MSA BW is increased, while employing resonant slot cut within the patch boundaries, which maintains the low profile of the antenna [4]. The slot cut at an appropriate location on the patch reduces the frequency of higher order mode closer to the fundamental mode, thus realizing wider BW. In addition to being compact in size and having an improved BW, antennas that can receive signals from any direction are preferable for wireless and personal communication applications because they reduce signal loss due to the multi-path propagation effects. Antennas offering higher cross-polar component of radiation (elliptical polarization) or those with a circular polarized response will be preferable in this situation.

This chapter presents ANN modelling of various compact slot cut RMSAs, SMSAs, and SCMSA for wider BW and CP responses. Initially, each MSA's working is explained with respect to the resonant modes and orthogonal modes resulting in a wider BW and CP response, followed by their ANN modelling.

5.2 ANN Model of Compact Slot Cut MSAs for Wideband and CP Response

5.2.1 Half E-Shape SMSA for Broadband Response

A half E-shape MSA is realized by using the even mode symmetry of E-shape MSA across the feed point axis [5]. The reported half E-shape MSA has simulated BW of 133 MHz (15.2%) against the 127 MHz (14.6%) of measured BW, with a peak broadside gain of 6 dBi over the reflection coefficient (S_{11}) BW. To understand the working of the antenna with respect to the resonant modes, the antenna with the same dimensions as reported in [5] is simulated. Initially, equivalent RMSA is simulated using IE3D software. The resonance frequency plot exhibits two peaks, and on studying the surface current distributions, those two modes are identified as TM_{10} and TM_{01} as shown in Fig. 5.1c, d. Here, as per the feed point location these two modes are excited, and the frequency of TM_{10} mode is smaller since $L > W$.

Further, the single rectangular slot is introduced as per the dimensions, and location given in [5]. To understand the modal behaviour the slot length, 'l_s' is parametrically varied. With the variation in slot length, reduction in TM_{01} mode frequency is observed, whereas TM_{10} mode frequency remains unchanged. This is because of the position of the slot in orthogonal direction to the TM_{01} mode surface currents. At $l_s = 8$ cm, and for the location of the slot and given x_f, as per [5] the optimum response is observed. To optimize impedance loop inside VSWR = 2 circle, the slot position is kept near the feed point axis [5]. The simulated impedance plot showing

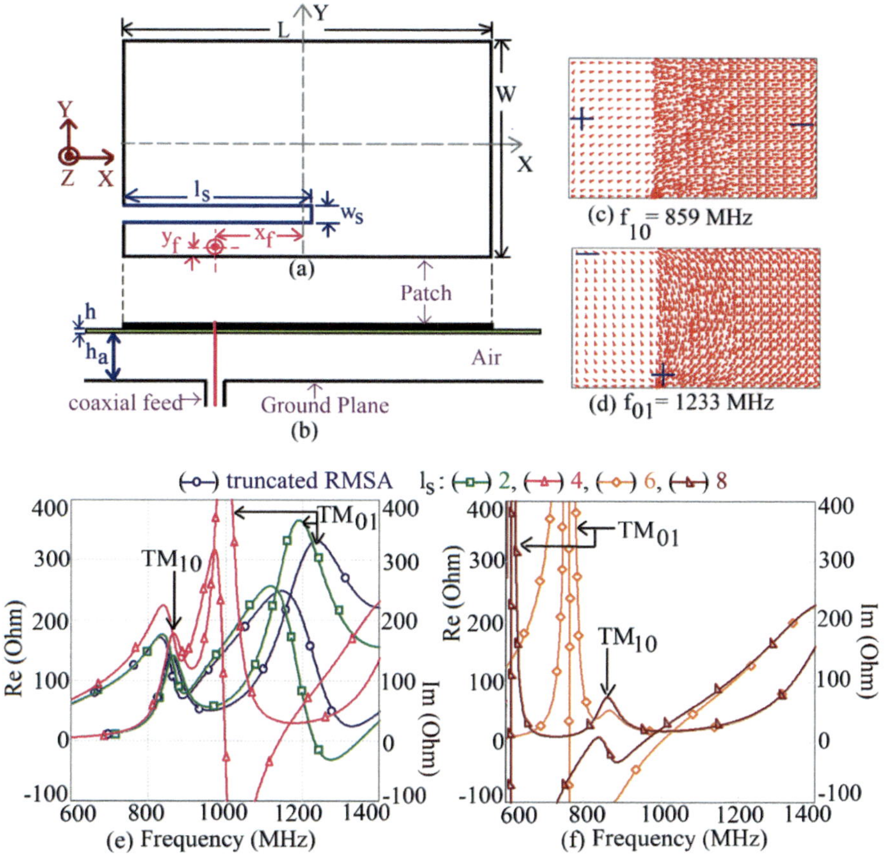

Fig. 5.1 **a**, **b** Air-suspended half E-shape MSA, its **c**, **d** surface current distributions, and **e**, **f** resonance curve plots for variation in slot length 'l_s'

the BW obtained is provided in Fig. 5.2. The MSA offers $S_{11} \leq -10$ dB BW of 122 MHz (13.96%), which is same as that obtained for the design in [5].

Further, based on the above analysis, ANN modelling of half E-shape MSA is proposed. Here air is used as the substrate. To propose ANN model for half E-shape MSA, it is crucial to defining the training data set for the network. For this initially, the half E-shape MSA is designed and optimized at various frequencies and substrate thicknesses. The half E-shape MSA design is carried out in the following manner for each frequency and for each substrate thickness to create training data set for the ANN, and the detail method of data set creation is represented in the flowchart as shown in Fig. 5.3.

1. Design RMSA for a particular TM_{10} mode frequency f_r, and substrate thickness h using resonant frequency equation of RMSA [2].

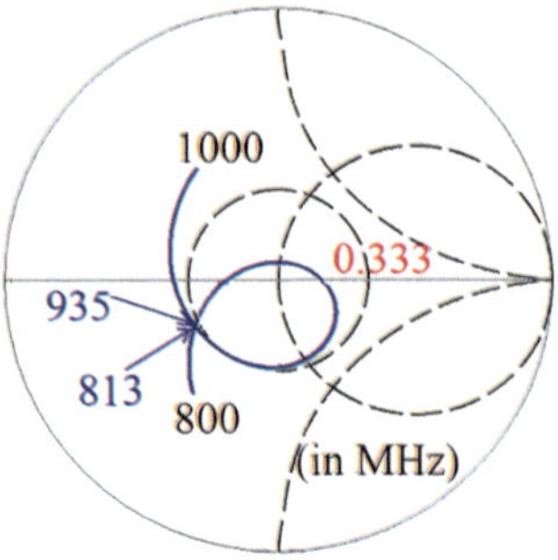

Fig. 5.2 Smith chart of wideband half E-shape MSA [5]

2. From the dimensions obtained in Step 1 to design half E-MSA, choose RMSA dimensions as L, $W/2$.
3. Feed the RMSA at appropriate location x_f, as shown in Fig. 5.1a, which initially exhibits two resonant peaks TM_{10} and TM_{01}.

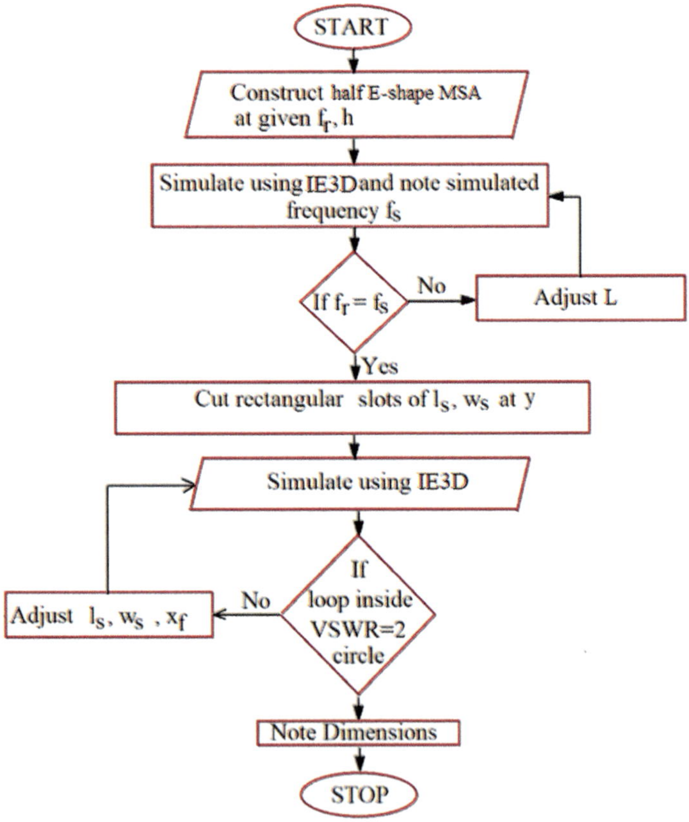

Fig. 5.3 Flowchart for wideband half E-shape MSA design for training data set generation

4. Further, optimize the slot dimensions l_s, w_s so that inter-spacing between TM_{10} and TM_{01} modes get tuned realizing wideband response by optimizing VSWR = 2 circle.

For training purpose of the neural network, simulations were carried out using IE3D Software over a frequency range of 800–5000 MHz in a step size of 200 MHz. For each frequency antenna, designs were simulated at four different substrate heights, i.e., 0.05λ–0.08λ. In total there were 112 points of data to train the neural network model. Sample training data set created using the above procedure is given in Table 5.1. All the dimensions in the table are expressed in cm, and frequencies are in MHz.

The block diagram and detailed structure of ANN model for the trained data are similar to the one shown in Fig. 2.4a, b with a modification in the suitable parameters. It is a deep neural network model with two hidden layers and ten neurons in each hidden layer. The optimum number of layers and number of neurons can only be decided by iterative testing of the result of the ANN. The number of layers and number of neurons in each layer for an iteration which have the least error and highest accuracy must be preferred. The architecture shown in Fig. 2.4a, b is designed as two parts. These are two ANNs communicating with each other to provide accurate predictions. The details of the ANN architecture such as number of inputs, outputs, and hidden layers are specified in Table 2.2. Part 1 of the architecture predicts an edge extension length quantity that accounts for fringing field extension. In this model, the Part 2 of the architecture predicts four antenna dimensions, l_s, w_s, y, w_s/L, and the feed point location x_f. Here y_f is considered as 0.16 cm from the lower edge of SMSA for designs under 2000 MHz, and 0.06 for the half E-shape MSA designs for frequency more than 2000 MHz.

Further, the trained ANN structure is used to predict the dimensions of half E-shape MSA for a given TM_{10} mode frequency (f_r) and substrate thickness (h_a). The dimensions predicted by the two-step ANN architecture is given in Table 5.2, after minor changes. These minor changes in the dimensions are done so as to obtain practically realizable values. From the dimensions observed from the table, the half E-shape MSA is designed, and simulations are conducted to verify the result in terms of resonance frequency peaks and S_{11} BW. For half E-shape MSA designed on air substrate at 750 MHz for $h_a = 0.08\lambda$, the results are shown in Fig. 5.4a. Simulated and measured S_{11} BWs of 184 MHz (23.23%), and 206 MHz (25.84%) are realized with a peak broadside gain of 8.1 dBi. The fabricated antenna using air substrate is shown in Fig. 5.4b.

5.2.2 *Half U-Slot Cut RMSA for Broadband Response*

Using the even mode symmetry of U-slot cut RMSA with respect to the feed point axis, and using only half of the geometry, a configuration of half U-slot cut RMSA is realized. The configuration of half U-slot RMSA as reported in [6] is shown in

Table 5.1 Sample ANN training data set of air-suspended half E-shape MSA

f_r	h_a	l	w	l_s	w_s	y	x_f
800	1.9	15.6	18.8	18.0	1.0	5.2	3.9
	2.2	15.0	18.0	18.4	1.0	3.6	3.7
	2.6	14.8	17.8	23.0	1.0	2.3	3.7
	3.0	14.4	17.4	27.2	1.0	1.4	3.6
900	2.0	13.5	16.2	17.0	1.0	3.2	3.4
	2.3	13.1	15.8	17.0	1.0	3	3.3
	2.7	12.8	15.4	20.0	1.0	2	3.2
1000	1.5	12.5	15.0	13.0	0.5	4.4	3.1
	1.8	12.1	14.6	15.0	1.0	3	3
	2.1	11.8	14.2	16.0	1.0	2.5	2.9
	2.4	11.5	13.8	18.8	1.0	1.8	2.9
1100	1.4	11.2	13.5	11.8	0.3	3.9	2.8
	1.6	11.0	13.3	12.2	1.0	3.4	2.7
	1.9	10.7	12.9	14.0	1.0	2.5	2.7
	2.2	10.5	12.7	18.4	1.0	1.5	2.6
1200	1.3	10.3	12.4	10.4	1.0	4.1	2.6
	1.5	10.1	12.1	11.8	1.0	3	2.5
	1.7	9.9	11.9	12.8	1.0	2.5	2.5
	1.9	9.7	11.7	13.4	1.0	2.2	2.4
1300	1.1	9.6	11.5	9.8	0.2	4.2	2.4
	1.4	9.3	11.2	11.0	0.8	2.7	2.3
	1.6	9.1	10.9	11.2	0.8	2.5	2.3
	1.8	8.9	10.7	13.0	0.8	1.8	2.2
1400	1.1	8.8	10.6	8.8	0.6	3.5	2.2
	1.3	8.6	10.4	10.0	0.8	2.5	2.1
	1.5	8.4	10.1	12.2	0.5	1.7	2.1
	1.7	8.2	9.9	15.0	0.6	1.1	2
1500	1.0	8.2	9.8	8.2	0.5	3.5	2
	1.2	8.0	9.6	9.0	0.8	2.5	2
	1.4	7.8	9.4	11.0	0.8	1.8	1.9
1600	0.9	7.8	9.4	7.7	0.4	3.7	1.9
	1.3	7.4	8.9	10.0	0.5	1.7	1.8
	1.5	7.2	8.7	11.6	0.5	1.3	1.8
1700	0.9	7.2	8.7	7.4	0.5	3.3	1.8
	1.0	7.1	8.5	7.6	0.5	2.8	1.8
	1.2	6.9	8.3	8.6	0.5	2	1.7
	1.4	6.7	8.1	11.0	0.5	1.3	1.7

(continued)

Table 5.1 (continued)

f_r	h_a	l	w	l_s	w_s	y	x_f
1800	0.8	6.9	8.3	7.0	0.5	3.5	1.7
	1.0	6.7	8.1	7.4	0.5	2.5	1.7
	1.2	6.5	7.8	9.0	0.5	1.6	1.6
	1.4	6.3	7.6	10.0	0.5	1.3	1.6
1900	0.8	6.5	7.8	6.6	0.3	3	1.6
	0.9	6.4	7.7	6.8	0.5	2.5	1.6
	1.1	6.2	7.5	7.8	0.5	1.8	1.5
	1.3	6.0	7.2	9.4	0.5	1.3	1.5
2000	0.8	6.1	7.3	6.15	0.5	2.8	1.5
	0.9	6.0	7.2	6.4	0.5	2.4	1.5
	1.0	5.9	7.1	6.8	0.5	2	1.5
	1.1	5.8	7.0	9.0	0.5	1.3	1.4
2200	0.7	5.6	6.7	5.6	0.5	2	1.4
	0.8	5.5	6.6	6.1	0.5	1.6	1.4
	0.9	5.4	6.5	7.3	0.5	1.1	1.3
	1.1	5.2	6.2	9.0	0.5	0.7	1.3
2400	0.6	5.2	6.2	5.0	0.5	2.1	1.3
	0.7	5.1	6.1	5.6	0.5	1.5	1.3
	0.9	4.9	5.9	6.8	0.5	1	1.2
	1.0	4.8	5.8	8.0	0.5	0.7	1.2
2600	0.6	4.8	5.7	4.8	0.3	1.8	1.2
	0.7	4.6	5.5	5.2	0.3	1.3	1.1
	0.8	4.6	5.5	6.0	0.3	1	1.1
	0.9	4.4	5.3	6.4	0.3	0.8	1.1
2800	0.5	4.5	5.4	4.2	0.3	1.8	1.1
	0.6	4.4	5.3	4.7	0.3	1.4	1.1
	0.7	4.3	5.2	5.0	0.3	1.2	1.1
	0.8	4.2	5.1	6.6	0.3	0.7	1
3000	0.5	4.1	4.9	4.0	0.3	1.5	1
	0.6	4.0	4.8	4.4	0.3	1.2	1
	0.7	3.9	4.7	5.2	0.3	0.9	1
	0.8	3.8	4.6	6.4	0.3	0.6	0.9
3200	0.5	3.8	4.6	3.9	0.3	1.3	0.9
	0.6	3.7	4.5	4.4	0.3	1	0.9
	0.7	3.6	4.3	5.4	0.3	0.7	0.9
	0.8	3.5	4.2	5.6	0.3	0.6	0.9
3400	0.4	3.7	4.4	3.55	0.3	1.7	0.9

(continued)

Table 5.1 (continued)

f_r	h_a	l	w	l_s	w_s	y	x_f
	0.5	3.6	4.3	3.8	0.3	1.8	0.9
	0.6	3.5	4.2	4.2	0.3	1	0.9
	0.7	3.4	4.1	5.4	0.3	0.6	0.8
3600	0.4	3.5	4.2	3.45	0.2	1.6	0.9
	0.5	3.4	4.1	4.0	0.2	0.9	0.8
	0.6	3.2	3.9	5.0	0.2	0.6	0.8
	0.7	3.1	3.7	4.8	0.2	0.6	0.85
3800	0.4	3.3	3.9	3.2	0.3	1.4	0.8
	0.5	3.1	3.7	3.6	0.3	0.9	0.8
	0.6	3.0	3.6	4.6	0.2	0.6	0.7
	0.7	3.0	3.6	5.4	0.2	0.4	0.7
4000	0.4	3.1	3.7	3.15	0.2	1.2	0.8
	0.5	3.0	3.6	3.7	0.2	0.8	0.7
	0.6	2.9	3.5	4.8	0.2	0.5	0.7
	0.7	2.8	3.4	5.2	0.2	0.4	0.7
4200	0.3	3.0	3.6	3.1	0.2	1.1	0.7
	0.4	2.9	3.5	3.1	0.2	1	0.7
	0.5	2.8	3.4	4.0	0.2	0.6	0.7
	0.6	2.7	3.3	4.0	0.2	0.6	0.7
4400	0.3	2.9	3.5	2.75	0.2	1.4	0.7
	0.4	2.7	3.2	2.85	0.2	1	0.7
	0.5	2.6	3.1	4.6	0.2	0.4	0.6
	0.6	2.6	3.1	5.0	0.2	0.3	0.6
4600	0.3	2.7	3.2	2.625	0.2	1.3	0.7
	0.4	2.6	3.1	2.8	0.2	1	0.7
	0.5	2.5	3.0	3.4	0.2	0.6	0.6
	0.6	2.4	2.9	4.2	0.2	0.4	0.6
4800	0.3	2.6	3.1	2.5	0.2	1.2	0.6
	0.4	2.5	3.0	2.8	0.2	0.9	0.6
	0.5	2.4	2.9	3.4	0.2	0.6	0.6
	0.6	2.3	2.8	4.4	0.1	0.3	0.6
5000	0.3	2.5	3.0	2.45	0.2	1.2	0.6
	0.4	2.4	2.9	2.8	0.2	0.8	0.6
	0.5	2.3	2.8	3.6	0.2	0.5	0.6
	0.6	2.2	2.7	4.3	0.2	0.3	0.5

Table 5.2 ANN predicted dimensions of air-suspended half E-shape MSA

f_r	h_a	L	W	l_s	w_s	y	x_f
750	3.2	15.5	18.599	23.966	1.207	2.25	3.953
950	1.9	12.857	15.429	14.614	1.053	3.696	3.237
950	2.5	12.169	14.602	18.708	0.943	1.835	3.045
1150	1.6	10.528	12.633	11.902	0.853	3.119	2.587
1150	2.1	9.936	11.923	15.245	0.766	1.539	2.441
1350	1.3	9.022	10.827	9.485	0.677	3.071	2.146
1350	1.8	8.402	10.082	12.862	0.645	1.329	2.03
1550	1.2	7.752	9.303	8.54	0.602	2.465	1.823
1550	1.5	7.37	8.844	11.012	0.571	1.244	1.746
1750	1	6.944	8.333	7.067	0.504	2.531	1.593
1750	1.4	6.427	7.713	9.88	0.493	1.017	1.46
1950	0.9	6.224	7.469	6.296	0.447	2.283	1.387
1950	1.2	5.83	6.996	8.722	0.45	0.976	1.257
2300	0.8	5.224	6.268	5.543	0.376	1.711	1.035
2300	1	4.963	5.955	7.347	0.386	0.838	0.938
2700	0.7	4.425	5.31	4.917	0.319	1.289	0.74
2700	0.9	4.173	5.007	6.616	0.33	0.615	0.708
3100	0.6	3.87	4.644	4.246	0.275	1.144	0.592
3100	0.8	3.623	4.347	5.881	0.289	0.531	0.594
3500	0.5	3.474	4.169	3.592	0.239	1.135	0.524
3500	0.7	3.227	3.872	5.168	0.257	0.514	0.522
3900	0.5	3.056	3.667	3.532	0.22	0.889	0.46
3900	0.6	2.935	3.523	4.415	0.229	0.564	0.466
4300	0.4	2.85	3.42	2.924	0.195	0.993	0.443
4300	0.6	2.613	3.136	4.348	0.208	0.436	0.442
4700	0.4	2.571	3.085	2.864	0.182	0.852	0.41
4700	0.5	2.452	2.943	3.643	0.189	0.55	0.405
5100	0.4	2.341	2.809	2.87	0.17	0.732	0.385
5100	0.5	2.232	2.679	3.595	0.175	0.453	0.387

Fig. 5.5a, b. As explained in [6] due to the reduction in patch width by half, mutual coupling between patch and the slot mode increases due to which the optimization of impedance loop inside VSWR = 2 circle becomes difficult [6]. Thus, to bring the loop inside VSWR = 2 circle, fringing fields can be enhanced to lower the radiation resistance [6]. This can be made possible by increasing the substrate thickness of the patch [6]. To understand the resonant mode behaviour of the half U-slot RMSA, antenna with the same dimensions as given in [6] is simulated using IE3D simulations.

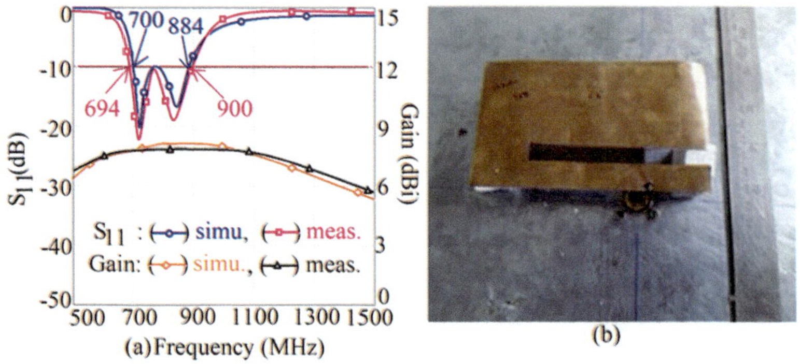

Fig. 5.4 **a** S_{11} BW and gain plots, and **b** fabricated prototype of half E-shape MSA

The initial resonant mode behaviour of equivalent RMSA for the specified feed point location is similar to the one shown in Fig. 5.1c–e. Further, while studying the surface current distributions at the resonance peaks observed in the resonance frequency plots, reduction in the TM_{10} and TM_{11} mode frequencies is observed against the increment in L_v. Similarly, increments in L_h reduce the frequencies of TM_{10}, TM_{01}, and TM_{11} resonant modes. For $L_h = 7.4$ cm, an optimum result in terms of impedance locus occupying inside VSWR = 2 circle is obtained. The surface current distributions are shown in Fig. 5.5c, d, and the simulated results are shown in Fig. 5.5e. A simulated S_{11} BW of 109 MHz (12%) with a peak gain of 8.1 dBi is realized, which is close to the results specified in [6]. Against RMSA loaded with U-slot, half U-slot RMSA realizes broadband response with a 50% reduction in the patch size.

Further, based on the working understood from the above analysis, ANN modelling of half U-slot RMSA is proposed. Here air is used as the substrate. To propose ANN model for half U-slot MSA, it is crucial to defining the training data set for the ANN. For this initially, the half U-slot cut RMSA is designed and optimized at various frequencies and substrate thickness. The MSA design is carried out in the following manner at each frequency and for each substrate thickness, to create the training data set for the ANN and the detail method of data set creation is represented in the flowchart shown in Fig. 3.3.

1. Design RMSA for a particular TM_{10} mode frequency f_r, and substrate thickness h using resonance frequency equation of RMSA [2].
2. From the dimensions obtained in step 1, choose RMSA width as $W = 0.6$ L as obtained from the optimum configuration.
3. Feed the RMSA at appropriate location as shown in Fig. 5.5a, which initially exhibits two resonant peaks TM_{10} and TM_{01}.
4. Further, optimize the slot dimensions L_v, L_h, and w_s so that both modes shift to lower frequency region. The impedance tuning of TM_{10} and TM_{01} optimizes impedance loop inside VSWR = 2 circle to achieve wideband response.

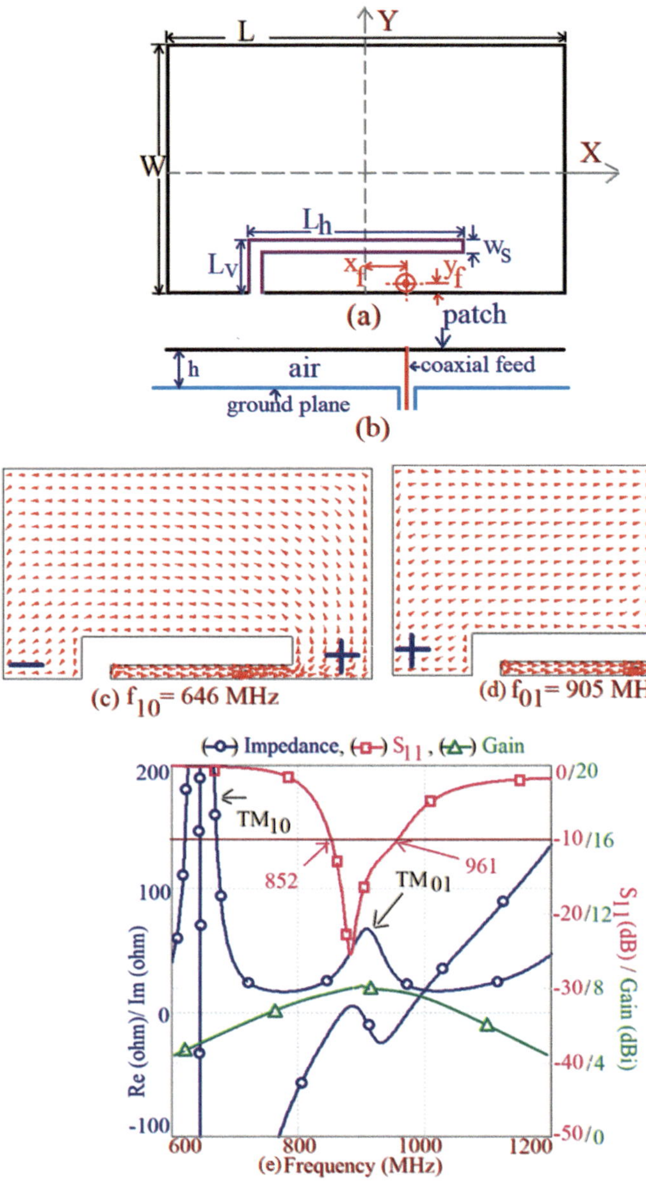

Fig. 5.5 **a**, **b** Half U-slot cut RMSA [6], its simulated **c**, **d** surface current distributions, and **e** resonance curve, S_{11} BW, and gain plot

Sample training data set of a single-stage ANN model created for the frequency range of 800–5000 MHz, on varying substrate thickness forms 0.05λ–0.1λ is shown in Table 5.3.

The procedure of ANN training is like the one adapted for U-slot cut RMSA discussed in Chap. 3. The block diagram and ANN nodal structure are like the ones shown for U-slot cut RMSA in Fig. 3.6. A two-stage neural network is designed to predict the dimensions of the half U-slot cut RMSA. In the two-stage neural network, two ANN models are designed, where the output of the first stage is given as an input to the second-stage network. The first neural network is trained to predict the patch dimensions of the RMSA. Hence, the inputs given to the first network are frequency and substrate thickness. The target data for this network is $(\Delta L)/h$. Thus, the value of $(\Delta L)/h$ is predicted for required frequencies and substrate thickness using this network. Here the length of the RMSA is calculated using $(\Delta L)/h$. Since the patch dimensions of the RMSA are obtained, these dimensions along with the frequency and substrate thickness are given as inputs to the second neural network. The target data given are the half U-slot parameters, i.e., L_h, L_v, w_s, and x_f. After training the second neural network on this data, the U-slot parameters are predicted for other frequency and substrate thicknesses. The result of this two-stage ANN is very efficient with the requirement of very fine optimization in some of the half U-slot parameters. Also, some defined optimization is done to generate practically realizable values for the MSA. The parameters of first layer ANN and second layer ANN in the double-stage model are given in Table 5.4.

To see the network's performance during training, performance graphs are plotted. Performance is measured in terms of mean squared error and is shown on a log scale. Mean squared error rapidly decreases as the network is trained. Performance is shown for each of the training, validation, and test sets. The final network is the network that performed best on the validation set. A training set is presented to the network during the training, and the network is adjusted according to its error. A validation set is used to measure network generalization and to halt training when generalization starts improving. Test data does not affect training and so provides an independent measure of network performance during and after training. The transgression graph for double-stage ANN using MSE function is plotted against the number of epochs used during the training algorithm. The results as displayed in Fig. 5.6a depict the Validation, Target, and Test Results. Mean squared error function is used to calculate the average squared difference between the targets and the outputs.

As observed, the lower the value of mean squared error, the more effective is the neural network is. Compared to the Single-Stage ANN network, the performance graph of the Double-Stage ANN network is observed to be exponentially increasing. The mean squared error is reduced by at least 450 units, which increases the accuracy of the prediction. Another measure of how well the neural network has fit the data is the regression plot. Here the regression is plotted across all samples as shown in Fig. 5.6b–e. The regression plot shows the actual network outputs plotted in terms of the associated target values. If the network has learned to fit the data well, the linear fit to this output-target relationship should closely intersect the bottom-left and top-right corners of the plot. If this is not the case then further training, or training

Table 5.3 Training data set of half U-slot MSA

f_r	$f_{simu.}$	h	$\Delta L/h$	L	W	L_h	L_v	w_s	x_f
800	805.6	2.25	0.08044	15	18	8	1.9	0.9	2.9
800	802.4	2.625	0.07607	14.7	17.64	9.2	2.2	0.9	1.7
800	800.8	3	0.07219	14.4	17.28	9.2	2.025	0.85	1.5
1000	1005.6	1.8	0.08101	12	14.4	6.45	1.575	0.7	2.3
1000	998.4	2.1	0.07676	11.8	14.16	6.7	1.575	0.7	2.1
1000	992.8	2.4	0.0731	11.6	13.92	6.9	1.75	0.7	1.5
1200	1204	1.5	0.08195	10	12	5.4	1.5	0.6	1.8
1200	1200	1.8	0.075	9.8	11.76	5.6	1.5	0.6	1.5
1200	1194.4	2	0.07397	9.6	11.52	5.8	1.5	0.6	1.2
1400	1400	1.3	0.08132	8.6	10.32	4.8	1.3	0.5	1.5
1400	1395.2	1.5	0.07837	8.4	10.08	5	1.35	0.5	1.2
1400	1395.4	1.7	0.07499	8.2	9.84	5	1.3	0.5	1.1
1600	1600	1.125	0.08333	7.5	9	3.9	1.15	0.45	1.5
1600	1600	1.3	0.07788	7.35	8.82	3.8	1.2	0.45	1.3
1600	1600	1.5	0.0725	7.2	8.64	4.4	1.2	0.45	1
1800	1824	1	0.08368	6.55	7.86	3.2	0.9	0.4	1.5
1800	1800	1.167	0.08283	6.4	7.68	3.9	1.1	0.4	1
1800	1800	1.333	0.07627	6.3	7.56	3.8	1.125	0.4	0.8
2000	2020	0.9	0.08476	5.9	7.08	3.5	0.95	0.4	1.1
2000	2020.8	1.05	0.07966	5.75	6.9	3.3	0.95	0.4	1.1
2000	1988	1.2	0.07689	5.7	6.84	3.4	1.1	0.4	0.8
2200	2195.7	0.818	0.08139	5.5	6.6	2.8	0.625	0.35	1.3
2200	2186.45	0.955	0.07646	5.4	6.48	3	0.7	0.35	1
2200	2189.8	1.091	0.07562	5.2	6.24	3	0.75	0.3	0.8
2400	2389.09	0.75	0.0819	5.05	6.06	2.7	0.575	0.3	1.1
2400	2382.48	0.875	0.07977	4.9	5.88	2.8	0.65	0.3	0.9
2400	2408.93	1	0.07634	4.7	5.64	2.55	0.6	0.3	0.9
2600	2616.86	0.692	0.0818	4.6	5.52	2.45	0.55	0.3	1
2600	2585.12	0.808	0.0806	4.5	5.4	2.6	0.625	0.3	0.8
2600	2624.13	0.923	0.07672	4.3	5.16	2.45	0.65	0.25	0.6
2800	2800	0.643	0.0822	4.3	5.16	2.2	0.575	0.25	0.9
2800	2795.04	0.75	0.08111	4.15	4.98	2.3	0.65	0.25	0.7
2800	2815.54	0.857	0.07745	4	4.8	2.4	0.675	0.25	0.5
3000	3000	0.6	0.08333	4	4.8	2.2	0.525	0.25	0.8
3000	3000	0.7	0.08214	3.85	4.62	2.2	0.575	0.2	0.55
3000	2995.7	0.8	0.07857	3.75	4.5	2.3	0.6	0.2	0.35
3200	3210	0.563	0.08933	3.667	4.4	1.875	0.46	0.2	0.7

(continued)

Table 5.3 (continued)

f_r	$f_{simu.}$	h	$\Delta L/h$	L	W	L_h	L_v	w_s	x_f
3200	3200	0.656	0.08418	3.583	4.3	2.15	0.55	0.2	0.5
3200	3212	0.75	0.0808	3.458	4.15	2.08	0.58	0.2	0.4
3400	3400	0.529	0.09015	3.458	4.15	1.92	0.465	0.2	0.65
3400	3400	0.618	0.08388	3.375	4.05	2	0.575	0.2	0.45
3400	3387	0.706	0.08057	3.291	3.949	2.1	0.55	0.2	0.35
3600	3558	0.5	0.09249	3.291	3.949	1.92	0.5	0.2	0.55
3600	3600	0.6	0.08331	3.167	3.8	2	0.5	0.2	0.4
3600	3600	0.667	0.08123	3.083	3.7	2.14	0.55	0.2	0.2
3800	3800	0.473	0.079	3.2	3.84	2	0.51	0.2	0.4
3800	3800	0.553	0.07662	3.1	3.72	2	0.55	0.2	0.35
3800	3827	0.632	0.07275	3	3.6	2	0.55	0.2	0.25
4000	4036	0.45	0.07962	3	3.6	1.6	0.5	0.2	0.6
4000	4038	0.525	0.07759	2.9	3.48	2	0.5	0.2	0.3
4000	4072	0.6	0.07364	2.8	3.36	2	0.5	0.2	0.2
4200	4226	0.43	0.08133	2.85	3.42	1.4	0.4	0.15	0.6
4200	4200	0.5	0.07964	2.775	3.33	1.4	0.425	0.15	0.5
4200	4206	0.57	0.0738	2.725	3.27	1.5	0.45	0.15	0.4
4400	4401	0.409	0.08292	2.73	3.276	1.2	0.4664	0.15	0.5
4400	4406	0.477	0.07908	2.65	3.18	1.3	0.477	0.15	0.4
4400	4400	0.545	0.07515	2.59	3.108	1.504	0.5162	0.15	0.3
4600	4620	0.391	0.08271	2.6	3.12	1.4	0.5	0.15	0.4
4600	4600	0.456	0.07795	2.55	3.06	1.5	0.6	0.15	0.3
4600	4627	0.521	0.07407	2.47	2.964	1.7	0.6	0.15	0.15
4800	4810	0.375	0.08247	2.5	3	1.2	0.375	0.15	0.55
4800	4770	0.4375	0.07825	2.46	2.952	1.4	0.475	0.15	0.4
4800	4772.56	0.5	0.0743	2.4	2.88	1.35	0.5	0.15	0.35
5000	5075	0.36	0.07718	2.4	2.88	1.2	0.3	0.15	0.55
5000	5000	0.42	0.07857	2.34	2.808	1.304	0.4712	0.15	0.4
5000	4975	0.48	0.07449	2.3	2.76	1.4	0.5	0.15	0.3

Table 5.4 Model parameters in both stages of two-step ANN model for half U-slot MSA

S. No.	Parameter	Values in first stage	Values in second stage
1	Number of inputs	3	4
2	Number of outputs	1	4
3	Number of hidden neurons	10	10
4	Number of epochs	186	2000
5	Training algorithm	LM	LM

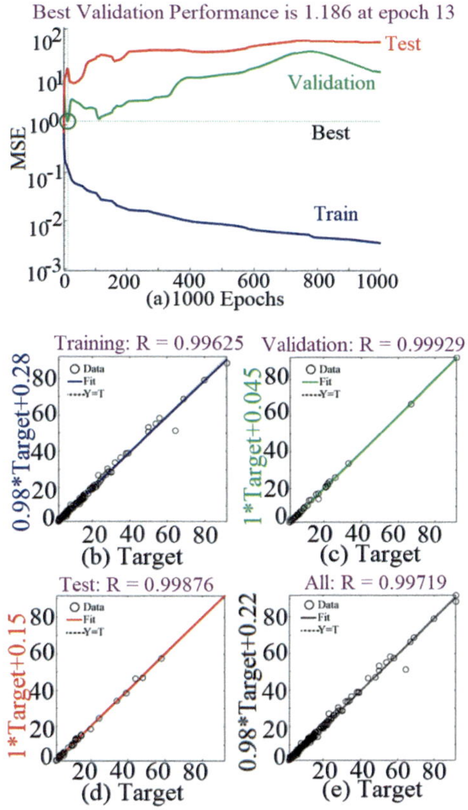

Fig. 5.6 **a** Performance graph, and **b**–**e** regression graphs of predicted values of double-stage ANN model of half U-slot MSA

a network with more hidden neurons, is done. The dimensions predicted by the double-stage ANN architecture for different frequencies and substrate thicknesses is given in Table 5.5, after minor changes.

These minor changes in the dimensions are done to obtain practically realizable values. The predicted data dimensions of half U-slot cut RMSA design at 1500 MHz on $h = 0.08\lambda$ are considered and optimized to get practically realizable values. The optimized and practically realizable dimensions of half U-slot cut RMSA are $L =$ 7.6 cm, $W = 4.6$, $h = 1.6$, $L_h = 4.7$, $L_v = 1.35$, $w_s = 0.5$, and $x_f = 1.1$ cm.

The simulated and measured S_{11} BW and gain plots are shown in Fig. 5.7a. A simulated BW of 276 MHz (16.7%), and measured BW of 300 MHz (18.18%) with a peak gain of 8 dBi are realized. The fabricated prototype of half U-slot cut RMSA at 1500 MHz on a substrate thickness of 0.08λ is shown in Fig. 5.7b.

Table 5.5 Predicted dimensions of half U-slot cut RMSA by double-stage ANN model

f_r	h	$\Delta L/h$	L	W	l_h	L_v	X_f	W_s
600	3.000	0.083	9.124	2.085	2.872	0.897	20.016	24.019
600	3.500	0.088	8.532	1.804	2.857	0.897	18.870	22.644
600	4.000	0.092	7.902	1.775	2.486	0.888	17.639	21.167
900	3.500	0.088	8.532	1.804	2.857	0.897	18.870	22.644
900	2.333	0.076	8.947	2.056	1.934	0.850	13.121	15.745
900	3.500	0.088	8.532	1.804	2.857	0.897	18.870	22.644
1500	1.200	0.083	4.242	1.192	1.548	0.494	8.008	9.610
1500	3.500	0.088	8.532	1.804	2.857	0.897	18.870	22.644
1500	1.600	0.074	4.700	1.340	1.095	0.485	7.648	9.178
2100	3.500	0.088	8.532	1.804	2.857	0.897	18.870	22.644
2100	1.000	0.078	3.170	0.819	0.969	0.349	5.592	6.710
2100	3.500	0.088	8.532	1.804	2.857	0.897	18.870	22.644
2700	0.667	0.082	2.324	0.580	0.961	0.266	4.457	5.348
2700	3.500	0.088	8.532	1.804	2.857	0.897	18.870	22.644
2700	0.889	0.077	2.443	0.615	0.563	0.245	4.179	5.015
3300	3.500	0.088	8.532	1.804	2.857	0.897	18.870	22.644
3300	0.636	0.084	2.020	0.531	0.475	0.217	3.473	4.168
3300	3.500	0.088	8.532	1.804	2.857	0.897	18.870	22.644
3900	0.462	0.080	2.012	0.508	0.417	0.202	3.105	3.726
3900	3.500	0.088	8.532	1.804	2.857	0.897	18.870	22.644
3900	0.615	0.072	2.185	0.539	0.272	0.197	2.958	3.550
4300	3.500	0.088	8.532	1.804	2.857	0.897	18.870	22.644
4300	0.488	0.078	1.355	0.446	0.457	0.154	2.722	3.266
4300	3.500	0.088	8.532	1.804	2.857	0.897	18.870	22.644
4900	0.367	0.081	1.254	0.400	0.650	0.158	2.463	2.956
4900	3.500	0.088	8.532	1.804	2.857	0.897	18.870	22.644
4900	0.490	0.074	1.495	0.572	0.230	0.161	2.334	2.801
5300	3.500	0.088	8.532	1.804	2.857	0.897	18.870	22.644
5300	0.396	0.075	1.221	0.358	1.112	0.159	2.238	2.686
5300	3.500	0.088	8.532	1.804	2.857	0.897	18.870	22.644
5500	0.330	0.076	1.208	0.332	1.859	0.159	2.227	2.672
5500	3.500	0.088	8.532	1.804	2.857	0.897	18.870	22.644
5500	0.430	0.073	1.214	0.344	1.269	0.159	2.101	2.521
1200	3.500	0.088	8.532	1.804	2.857	0.897	18.870	22.644
1800	1.330	0.074	3.572	1.116	0.854	0.389	6.357	7.628

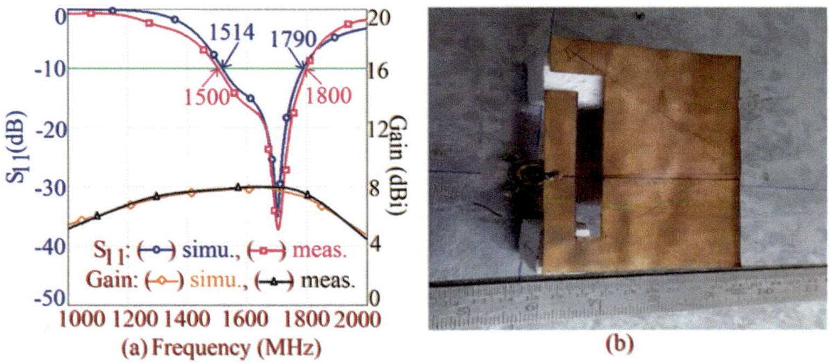

Fig. 5.7 **a** S_{11} BW and gain plots, and **b** fabricated prototype of half U-slot MSA designed at 1500 MHz on $h = 0.08\lambda$ using double-stage ANN model

5.2.3 Double Half U-Slot Cut RMSA for Broadband Response

The MSAs designed for BW enhancement exhibit linear polarization over the complete S_{11} BW, where cross-polar component lies about 15 dB below the co-polar component of radiation. This is due to their functioning around the fundamental mode supported with a modified surface current distribution at next higher order mode contributing to the S_{11} BW. However, MSAs should be able to receive signals from any polarization directions to avoid the signal loss due to the signal fading. Thus, enhanced levels of cross-polar components are required to fulfil this criterion. This can be realized by achieving bidirectional current flow on the patch aperture. A double half U-slot cut design as reported in [7] is designed to realize this high cross-polar component of radiation by creating the asymmetry in the two half-U-slot cut square MSA. A proximity feeding was considered to realize wider BW on thicker substrate. The structure of double half U-slot cut nearly SMSA reported in [7] is shown in Fig. 5.8.

The working of the antenna with respect to the resonant modes realizing wider BW and increased cross-polar component is explained in [7] in detail. Initially, without any slots, as the MSA is a nearly square patch, with respect to the given feed point location, no separate peaks for orthogonal TM_{10} and TM_{01} modes are observed. As the feed is at the offset location, TM_{11} modes gets excited. With the introduction of the first half U-slot, with variation in L_{v1}, TM_{10}, TM_{11}, TM_{20} mode frequencies decrease as the slot position is in orthogonal direction to the surface currents of those modes. Further, the increment in L_{h1} decreases frequency of all resonant modes. However, TM_{10} frequency shifts away from TM_{11} mode thus not causing wider BW. Improvement in the BW is achieved by tuning the next higher order TM_{20} mode frequency with the introduction of second half U-slot with dimensions L_{v2} and L_{h2}. The presence of modified TM_{11} and TM_{20} resonant modes leads to the bidirectional surface current variation over the patch. Although a broadside radiation pattern is

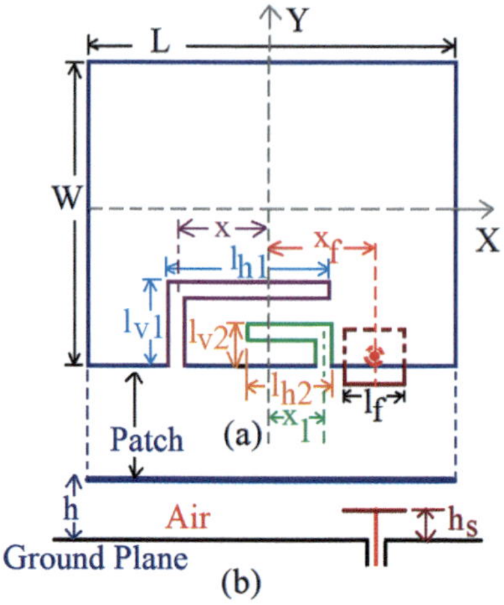

Fig. 5.8 Proximity-fed double half U-slot SMSA for broadband response with an increased cross-polar radiation [7]

observed over the BW, higher cross-polar component is also present which will be advantageous in mobile and personal communication systems. As entire working of the antenna is explained clearly in [7], the same details about resonant modes, radiation patterns, and surface current distributions are not mentioned here.

To propose the ANN model of the double half U-slot cut MSA, the training data set is created in the following manner:

1. Initially, square MSA is designed for its fundamental mode frequency of given f_r and *h*. The length of the MSA is calculated by using the equation given in [2], and the width of the patch is chosen as 0.95*L*, i.e., nearly square patch.
2. Further, length, width, and position of first half U-slot is optimized to bring the impedance loop inside VSWR = 2 circle, with the impedance tuning of TM_{11} mode.
3. The length, width, and position of second half U-slot is optimized to tune TM_{20} mode so that wider BW with increased cross polarization is achieved.

The above procedure is repeated for frequencies in the range of 800–5000 MHz in a step size of 200 MHz for values of '*h*' ranging from 0.05λ to 0.1λ. A sample training data set created using the above procedure is given in Table 5.6.

The procedure of creating training data set and training ANN model is shown in the flowchart given in Fig. 5.9. It can be seen from the flowchart that a two-stage neural network is designed to predict the dimensions of both the half U-slots in nearly square MSA. In the two-stage neural network, two ANN models are designed, where the output of the first stage is given as an input to the second-stage network. The first neural network is trained to predict the patch dimensions of the square MSA. Hence, the inputs given to the first network are resonant frequency 'f_r' and substrate thickness '*h*'. The target data for this network is $(\Delta L)/h$. Thus, the value of $(\Delta L)/h$ is predicted for required frequencies and substrate thickness using this network. Here the length and width of the nearly square MSA are calculated using $(\Delta L)/h$.

Table 5.6 ANN training data set of double U-slot cut MSA

f_r	h	$\Delta L/h$	L	W	l_{v1}	l_{h1}	l_{v2}	l_{h2}	w_s	x, x_1	l_f	x_f	h_s
800	3.75	0.07	13.9	13.2	4.02	8.74	1.78	3.58	0.8	4	2.25	4.6	3.3
1000	3	0.07	11	10.4	3.17	6.9	1.4	2.8	0.6	3.2	1.8	3.43	2.5
1200	2.5	0.07	9.2	8.76	2.67	5.8	1.18	2.4	0.55	2.7	1.7	3.04	2
1400	2.14	0.07	7.8	7.4	2.25	4.9	1	2	0.45	2.3	1.4	2.57	1.8
1600	1.88	0.07	6.8	6.5	2	4.3	0.9	1.8	0.4	2.2	1.225	2.24	1.65
1800	1.67	0.07	6	5.7	1.74	3.77	0.77	1.55	0.36	1.74	1.2	1.83	1.4
2000	1.5	0.06	5.6	5.3	1.6	3	0.7	1.2	0.3	1.6	1	1.8	1.3
2200	1.36	0.07	5	4.76	1.45	3.15	0.64	1.3	0.3	1.45	0.72	1.65	1.227
2400	1.25	0.07	4.6	4.38	1.335	2.9	0.6	1.18	0.3	1.33	0.65	1.46	1.125
2600	1.15	0.07	4.25	4.048	1.235	2.673	0.546	1.0965	0.25	1.2325	0.6	1.35	1.038
2800	1.07	0.07	3.95	3.762	1.147	2.485	0.507	1.019	0.25	1.145	0.543	1.2535	0.964
3000	1	0.07	3.65	3.476	1.06	2.3	0.469	0.9417	0.2	1.0585	0.5	1.155	0.9
3200	0.94	0.07	3.45	3.28	1	2	0.44	0.89	0.2	1.11	0.56	1.14	0.84
3400	0.88	0.07	3.25	3.09	0.94	2.04	0.42	0.84	0.2	0.94	0.62	1.0725	0.7
3600	0.83	0.07	3.05	2.9	0.88	1.92	0.39	0.79	0.2	0.88	0.6	1	0.67
3800	0.79	0.07	2.85	2.7	0.82	1.79	0.36	0.74	0.2	0.83	0.57	0.94	0.62
4000	0.75	0.07	2.7	2.57	0.78	1.69	0.35	0.69	0.2	0.78	0.55	0.89	0.6
4200	0.714	0.07	2.6	2.46	0.75	1.64	0.33	0.67	0.15	0.75	0.51	0.86	0.6
4400	0.68	0.07	2.5	2.37	0.72	1.57	0.32	0.65	0.15	0.725	0.5	0.825	0.58
4600	0.65	0.07	2.4	2.27	0.69	1.51	0.31	0.62	0.15	0.69	0.48	0.79	0.55
4800	0.63	0.07	2.3	2.18	0.66	1.45	0.29	0.59	0.15	0.67	0.465	0.76	0.53
5000	0.6	0.077	2.2	2.08	0.64	1.38	0.28	0.57	0.15	0.64	0.45	0.726	0.5

Since the patch dimensions of the square patch are obtained, these dimensions along with the frequency and substrate thickness are given as inputs to the second neural network. The target data given is double half U-Slot parameters, i.e., horizontal slot lengths for the larger and smaller half U-slot, vertical slot lengths for the larger and smaller half U-slot, half U-slots positions, slot width, proximity feed strip length and position, and proximity strip height. After training the second neural network on this data, the half U-slot parameters are predicted for the other frequency and substrate thicknesses. The result of this two-stage ANN is very efficient with the requirement of very fine optimization in some of the half U-slot parameters. Also, some defined optimization is done to generate practically realizable values for the antenna. The neural network model used for the design of half U-slots cut nearly square MSA is shown in Fig. 5.10.

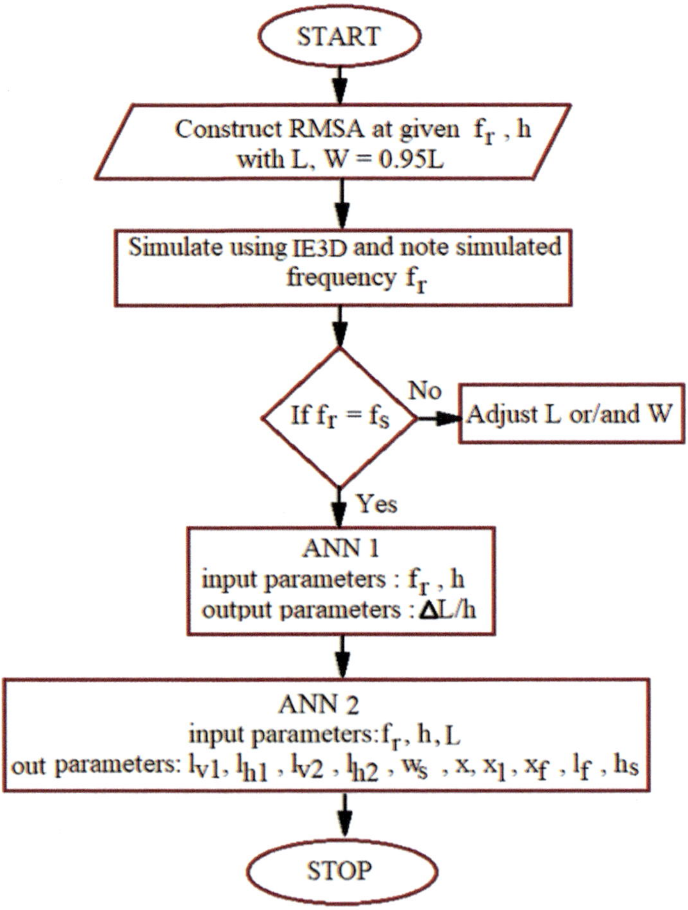

Fig. 5.9 Flowchart of two-stage ANN model of double half U-slot MSA

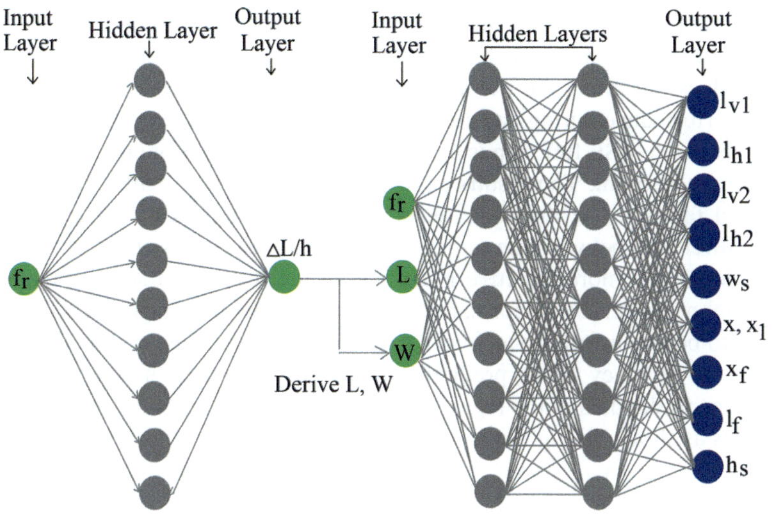

Fig. 5.10 ANN model with all layers of two-stage ANN for double half U-slot cut MSA

The details of the ANN architecture such as number of inputs, outputs, and hidden layers are specified in Table 5.7. The number of inputs, outputs, and hidden neurons are fed to the neural network in accordance with the training needed. The number of epochs is set as high as possible to terminate the training based on the error rates generated. The LM, also known as DLS method, is used to solve nonlinear least square problems. These minimization problems arise, especially in the least square curve fitting.

The data set is automatically divided into training, validation, and test sets when the network is ready to be trained. The training set is used to teach the network. Training continues as long as the network continues improving on the validation set. The test set provides a completely independent measure of network accuracy. To see the network's performance during training, a performance graph is plotted. Performance is measured in terms of mean squared error and shown on a log scale. It rapidly decreases as the network is trained. Performance is shown for each of the training, validation, and test sets. The final network is the network that performed best on the validation set. The training set is presented to the network during training

Table 5.7 Model parameters in both stages of two-step ANN model for double half U-slot cut MSA

S. No.	Parameter	Values in first stage	Values in second stage
1	Number of inputs	3	5
2	Number of outputs	1	9
3	Number of hidden neurons	10	10
4	Number of epochs	186	2000
5	Training algorithm	LM	LM

and the network is adjusted according to its error. A validation set is used to measure network generalization and to halt training when generalization starts improving. Test data does not affect the training and provides an independent measure of the network performance during and after the training. The transgression graph for two-stage ANN using MSE function is plotted against the number of epochs used during the training algorithm. The results as displayed in Fig. 5.11 depict the Validation, Target, and Test Results. Mean squared error function is used to calculate the average squared difference between the targets and the outputs. As observed, the lower the value of MSE, the more effective is the neural network. Compared to the Single-Stage ANN network, the performance graph of the Double-Stage ANN network is observed to be exponentially increasing. The mean squared error is reduced, which increases the accuracy of the prediction. Another measure of how well the neural network has fit the data is the regression plot. Here the regression is plotted across all samples as shown in Fig. 5.12a–d.

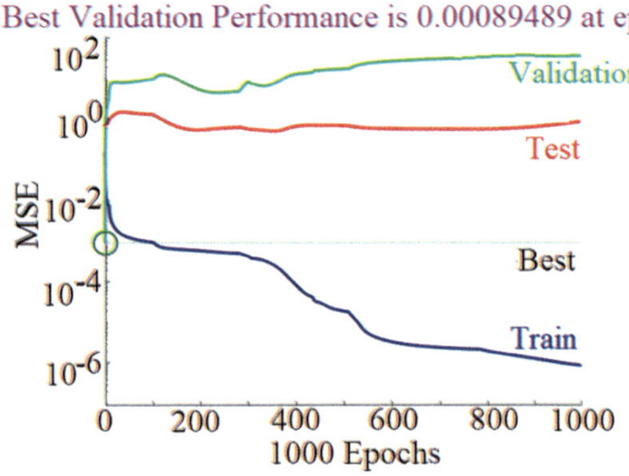

Fig. 5.11 Performance graph of two-stage ANN model of double half U-slot cut MSA

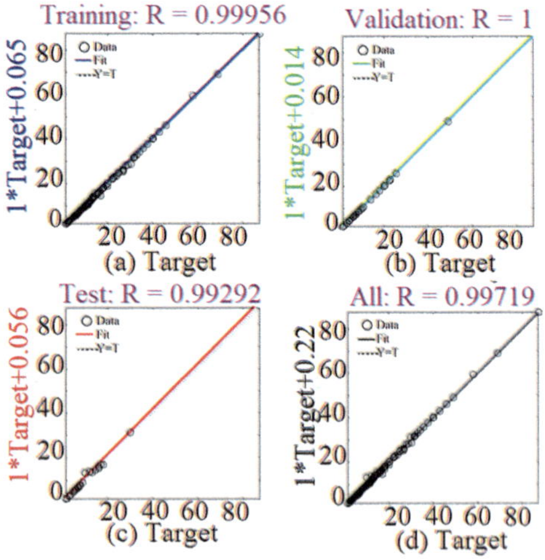

Fig. 5.12 **a–d** Regression graphs of predicted values of two-stage ANN model of double half U-slot MSA

Table 5.8 ANN predicted dimensions of double half U-slot cut MSA

f_r	h	$\Delta L/h$	L	W	l_{v1}	l_{h1}	l_{v2}	l_{h2}	w_s	x, x_1	l_f	x_f	h_s
600	5	0.065	18.4	17.5	4	8.7	1.8	3.6	0.8	4	2.5	4.6	3.3
900	3.3	0.065	12.3	11.7	4	8.7	1.8	3.6	0.8	4.9	2.25	4.6	2.9
1300	2.3	0.066	8.5	8.1	2.3	5.7	1.05	2.4	0.45	2.3	1.15	2.85	1.85
1700	1.75	0.07	6.35	6.05	1.8	3.9	0.8	1.65	0.35	2	1.05	2.1	1.5
2300	1.3	0.066	4.8	4.55	1.4	3	0.6	1.2	0.3	1.4	0.7	1.55	1.15
2700	1.1	0.06	4.1	3.9	1.2	2.6	0.55	1.05	0.25	1.2	0.55	1.3	1
3300	0.9	0.066	3.35	3.2	0.95	2.05	0.45	0.85	0.2	1	0.6	1.1	0.75
3700	0.8	0.067	2.95	2.8	0.85	1.85	0.4	0.75	0.2	0.85	0.6	0.95	0.65
4300	0.7	0.067	2.55	2.45	0.75	1.6	0.3	0.65	0.15	0.75	0.5	0.85	0.55
4700	0.65	0.065	2.35	2.25	0.7	1.45	0.3	0.6	0.15	0.65	0.45	0.75	0.55
5300	0.55	0.069	2.05	1.95	0.65	1.4	0.3	0.55	0.15	0.65	0.45	0.75	0.5
5700	0.5	0.07	1.9	1.8	0.65	1.4	0.3	0.55	0.15	0.65	0.45	0.75	0.5

The regression plot shows the actual network outputs plotted in terms of the associated target values. If the network has learned to fit the data well, the linear fit to this output-target relationship should closely intersect the bottom-left and top-right corners of the plot. If this is not the case then further training, or training a network with more hidden neurons, is done. Further, the trained ANN structure is used to predict the dimensions of half U-slot cut MSA for a given TM_{10} mode frequency (f_r) and substrate thickness (h). The dimensions predicted by the double-stage ANN architecture is given in Table 5.8, after minor changes. These minor changes in the dimensions are done to obtain practically realizable values.

From the predicted data, dimensions for two half U-slot cut nearly square MSA at 1700 MHz are considered and optimized to get practically realizable values. The practically realizable dimensions of half U-slot cut nearly square MSA are $L = 6.35$, $W = 6.05$, $h = 1.75$, $L_{v1} = 1.8$, $L_{h1} = 4.0$, $L_{v2} = 0.8$, $L_{h2} = 1.65$, $w_s = 0.35$, $x = x_1 = 1.825$, $l_f = 1.05$, $h_s = 1.75$, and $x_f = 2.1$ cm. The simulated and measured S_{11}, and gain plots are shown in Fig. 5.13a. Simulated BW of 1320 MHz (51.6%), and a measured BW of 1384 MHz (54.14%), with a peak gain of about 7.6 dBi is realized. The fabricated prototype of double half U-slot MSA at 1700 MHz on a substrate thickness of 0.09λ is shown in Fig. 5.13b.

5.2.4 *SCMSA with a Rectangular Slot*

A compact version of broadband CMSA with a pair of rectangular slots is the SCMSA loaded with a single rectangular slot. This configuration is obtained by using the symmetry of CMSA along the coaxial feed point axis. This broadband variation of

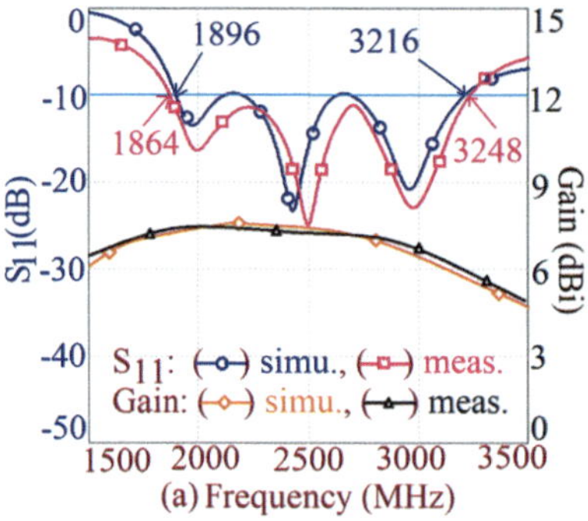

(b)

Fig. 5.13 **a** S_{11} BW and gain plots and **b** fabricated prototype for double half U-slot cut nearly square MSA designed at 1700 MHz, on 0.09λ substrate thickness

SCMSA with rectangular slot is reported in [8] and is as shown in Fig. 5.14a, b. To enhance the BW, the slot cut SCMSA is designed on air substrate as shown below.

As explained in [8], TM_{11} and TM_{21} modes are excited on SCMSA with respect to the given location of the feed point. Initially, with respect to the surface current distributions, the orientation of TM_{21} mode surface currents is mostly vertical, whereas that of TM_{11} mode is horizontal. Due to this, the radiation pattern exhibits E-plane along $\Phi = 90^0$ at the TM_{21} mode. With the increment in slot length, reduction in TM_{21} mode frequency is observed. This is due to the orthogonal orientation of TM_{21} mode surface currents against the slot length. This tuning of TM_{21} mode with respect to slot length re-orients the surface currents at that mode in horizontal direction. This changes E-plane direction from $\Phi = 90^0$ to 0^0 at the modified TM_{21} mode. The reduction of TM_{21} mode frequency nearer to the TM_{11} mode frequency widens the BW. As the analysis of the SCMSA with rectangular slot with respect to resonant mode, surface current distributions, and radiation patterns is given [8], the same is not repeated here.

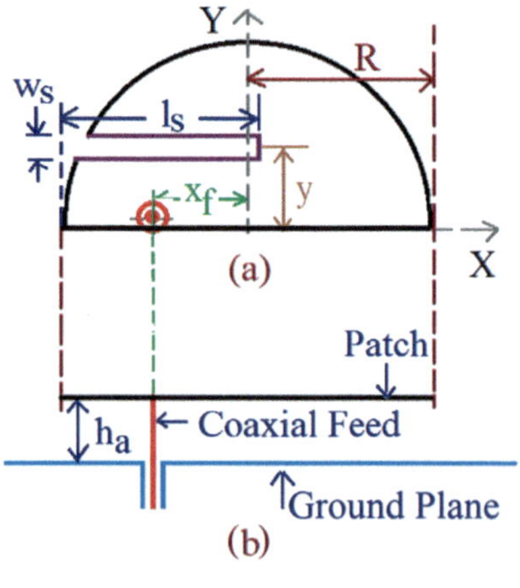

Fig. 5.14 **a, b** SCMSA with rectangular slot [8]

To design the ANN model, the crucial step is training the ANN through the data sets. For this initially, the SCMSA with a rectangular slot is optimized for broadband response by tuning the separation between TM_{11} and TM_{21} resonant mode frequencies over a frequency range of 800–5000 MHz in steps of 200 MHz. Here air is used as the substrate. At each frequency, SCMSA with a rectangular slot is optimized for substrate heights, varying in the range 0.05λ–0.08λ. The antenna design is carried out in the following manner for each frequency and for each substrate thickness to create a training data set for the ANN, and the detail method of data set creation is similar to the one given in the flowchart as shown in Fig. 3.2.

1. Design an SCMSA for its TM_{11} mode frequency at resonance frequency f_r using CMSA resonance frequency equation as given in [2].
2. Optimize the radius of the patch to get TM_{11} mode frequency at the required frequency.
3. Cut a rectangular slot at the maximum current location of TM_{21} mode, with respect to the simulated surface current distribution plot.
4. Optimize the slot length, width, and position to tune TM_{21} mode frequency nearer to the TM_{11} mode.
5. Optimize the feed point location to shift the impedance locus inside VSWR = 2 circle in the Smith chart.

The training data set created using the above procedure is given in Table 5.9. All the frequency and antenna dimensions in the table are in MHz and cm, respectively.

The two-stage ANN model of SCMSA with a rectangular slot is shown in Fig. 5.15a, b. The two-step ANN is a combination of two different ANN models. Instead of using multiple parameters in a single ANN, the first model in this method uses input frequency (f), half-wavelength ($\lambda/2$), dielectric constant (ε_r) of the substrate, substrate thickness (h), and $\Delta R/h$ value as input parameters. Radius (R) of the circular patch is the estimated output of the network. This optimized value of radius is then used as an input parameter in the second model of the process, along with slot length l_s, slot width w_s, slot separation y, feed position x_f, dielectric constant (ε_r), half-wavelength ($\lambda/2$), and the $\Delta r/h$ value.

The estimated outputs of the second model are x_f, h_a, l_s, w_s, y, and R. In both the models, we have used the LM algorithm for training the data. Both the models consist of 10 hidden neurons and were trained on 1000 epochs. The modal parameter tables in both the stages of two-step ANN model are like the ones given in Table 4.4 given in Chap. 4. The model accuracy can be increased by increasing the number of hidden neurons, but using too many hidden neurons may lead to overfitting of the model. For testing, data was taken from 900 MHz up to 5500 MHz in steps of 200 MHz. For each frequency, four different substrate heights were taken. These substrate heights (in terms of wavelength) were 0.05λ, 0.06λ, 0.07λ, and 0.08λ. For prediction, frequencies from 900 to 5100 MHz in steps of 600 MHz are considered. Further, the plot between the MSE and number of iterations is indicated as a performance plot. The best results for validation, testing, and training are plotted in this. MSE is the average squared difference between the targets and outputs. Lower the value of MSE, more is the effectiveness of the network. Performance plots for both the models are shown in

Table 5.9 Antenna training for SCMSA embedded with a rectangular slot

f_r	h	R	x_f	l_s	w_s	y	BW	$\Delta R/h$
800	2.3	9	− 3.3	9.4	0.8	2.6	0.006	0.0163
800	2.6	8.8	− 3.5	9.8	0.8	2.5	0.013	0.0221
800	3	8.5	− 3.6	10.1	0.8	2.4	0.017	0.0292
1000	1.8	7.2	− 1.9	5.5	0.8	2	0.012	0.0167
1000	2.1	7	− 2.7	7.6	0.7	2	0.014	0.0238
1000	2.4	6.8	− 2.7	8	0.7	1.9	0.022	0.0292
1200	1.5	5.95	− 1.5	4.5	0.6	1.7	0.015	0.02
1200	1.8	5.85	− 2.3	6.6	0.6	1.7	0.02	0.0222
1200	2	5.75	− 2.3	6.6	0.6	1.6	0.024	0.025
1400	1.3	5.12	− 1.3	3.9	0.6	1.4	0.018	0.0182
1400	1.5	5.02	− 1.8	5.3	0.6	1.4	0.014	0.0225
1400	1.7	4.93	− 1.9	5.7	0.6	1.3	0.03	0.0251
1600	1.1	4.46	− 1	3.3	0.5	1.3	0.022	0.0207
1600	1.3	4.37	− 1.5	4.8	0.5	1.3	0.018	0.0244
1600	1.5	4.18	− 1.5	5.2	0.5	1.2	0.036	0.0338
1800	1	4	− 0.9	3	0.5	1.1	0.019	0.0167
1800	1.2	3.86	− 1.2	4.3	0.5	1.1	0.019	0.0256
1800	1.3	3.72	− 1.3	4.3	0.5	1	0.038	0.0344
2000	0.9	3.55	− 0.6	2.7	0.4	1	0.029	0.0222
2000	1.1	3.48	− 1.1	3.8	0.4	1	0.022	0.0245
2000	1.2	3.35	− 1	4	0.4	0.9	0.038	0.0333
2200	0.82	3.2	− 0.9	2.4	0.6	0.8	0.025	0.0255
2200	0.95	3.2	− 0.8	2.45	0.6	0.8	0.025	0.022
2200	1.1	3	− 0.7	2.4	0.6	0.7	0.016	0.0372
2400	0.75	3	− 0.7	2.3	0.6	0.7	0.024	0.0167
2400	0.88	2.9	− 0.7	2.2	0.6	0.7	0.027	0.0256
2400	1	2.9	− 0.7	2.25	0.6	0.7	0.022	0.0225
2600	0.69	2.7	− 0.5	2.1	0.5	0.6	0.026	0.0268
2600	0.81	2.7	− 0.7	2	0.6	0.6	0.028	0.0228
2600	0.92	2.6	− 0.6	2	0.6	0.6	0.025	0.0309
2800	0.64	2.5	− 0.6	1.8	0.5	0.55	0.037	0.0279
2800	0.75	2.5	− 0.6	1.9	0.5	0.55	0.028	0.0238
2800	0.86	2.4	− 0.4	1.95	0.5	0.55	0.022	0.0324
3000	0.6	2.3	− 0.5	1.7	0.4	0.55	0.039	0.0333
3000	0.7	2.3	− 0.5	1.8	0.4	0.55	0.03	0.0286
3000	0.8	2.2	− 0.4	1.75	0.5	0.45	0.024	0.0375
3200	0.56	2.23	− 0.5	1.7	0.3	0.65	0.044	0.0202

(continued)

Table 5.9 (continued)

f_r	h	R	x_f	l_s	w_s	y	BW	$\Delta R/h$
3200	0.66	2.19	– 0.6	1.7	0.3	0.65	0.039	0.0234
3200	0.75	2.1	– 0.5	1.7	0.3	0.65	0.033	0.0325
3400	0.53	2.1	– 0.5	1.6	0.3	0.65	0.049	0.02
3400	0.62	2.06	– 0.5	1.6	0.3	0.55	0.039	0.0236
3400	0.71	2	– 0.6	1.7	0.2	0.6	0.027	0.0292
3600	0.5	1.98	– 0.7	1.4	0.2	0.8	0.067	0.0207
3600	0.58	1.94	– 0.7	1.5	0.2	0.85	0.06	0.0246
3600	0.67	1.88	– 0.6	1.5	0.2	0.8	0.054	0.0305
3800	0.47	1.88	– 0.6	1.4	0.2	0.7	0.05	0.0198
3800	0.55	1.82	– 0.5	1.4	0.2	0.7	0.058	0.0278
3800	0.63	1.78	– 0.6	1.5	0.2	0.9	0.063	0.0307
4000	0.45	1.78	– 0.6	1.4	0.2	0.8	0.054	0.0211
4000	0.53	1.72	– 0.6	1.4	0.2	0.8	0.062	0.0295
4000	0.6	1.68	– 0.6	1.35	0.2	0.8	0.069	0.0325
4200	0.43	1.65	– 0.6	1.3	0.2	0.8	0.066	0.0316
4200	0.5	1.6	– 0.6	1.3	0.2	0.8	0.075	0.0371
4200	0.57	1.55	– 0.7	1.2	0.2	0.8	0.111	0.0414
4400	0.41	1.6	– 0.6	1.3	0.2	0.8	0.073	0.0255
4400	0.48	1.6	– 0.6	1.3	0.2	0.8	0.084	0.0218
4400	0.55	1.55	– 0.6	1.3	0.2	0.8	0.082	0.0281
4600	0.39	1.55	– 0.6	1.3	0.2	0.8	0.06	0.0206
4600	0.45	1.5	– 0.7	1.2	0.2	0.8	0.097	0.029
4600	0.51	1.45	– 0.6	1.2	0.2	0.8	0.108	0.0354
4800	0.38	1.48	– 0.6	1.2	0.2	0.8	0.09	0.022
4800	0.44	1.42	– 0.6	1.2	0.2	0.8	0.104	0.0326
4800	0.5	1.4	– 0.6	1.1	0.2	0.7	0.115	0.0325
5000	0.36	1.42	– 0.5	1.1	0.2	0.7	0.102	0.0222
5000	0.42	1.36	– 0.5	1.1	0.2	0.7	0.103	0.0333
5000	0.48	1.34	– 0.5	1.1	0.2	0.7	0.11	0.0333

Fig. 5.16. For the first model, the best validation performance is 0.58106 at epoch 208 and for the second model, it is 0.42265 at epoch 38. The dimensions predicted by the two-step ANN architecture are given in Table 5.10, after minor changes. These minor changes in the dimensions are done so as to obtain practically realizable values.

Further, from the dimensions observed from the above table, the rectangular slot-loaded SCMSA is designed using air substrate for resonance frequency of 900 MHz on 0.07λ substrate thickness. The results in terms of S_{11} BW and gain are shown in Fig. 5.17a. Simulated and measured S_{11} BW of 53 MHz (5.9%), and 57 MHz

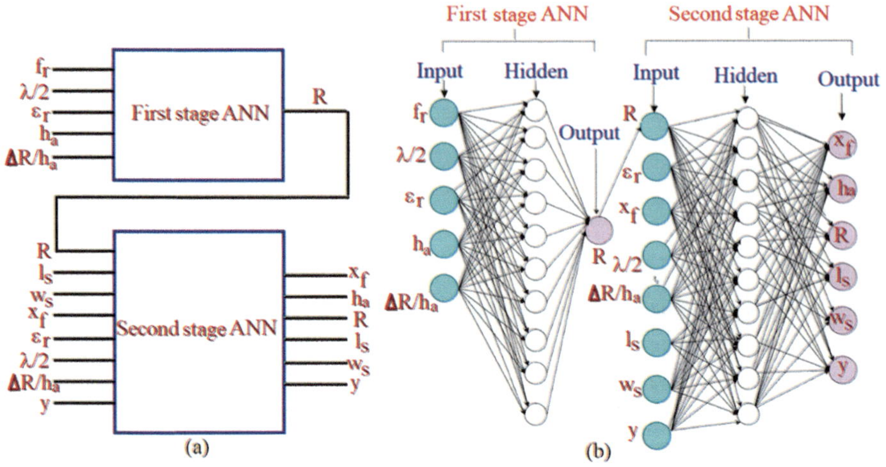

Fig. 5.15 **a** Block diagram and **b** nodal structure of two-stage ANN model of rectangular slot-loaded SCMSA

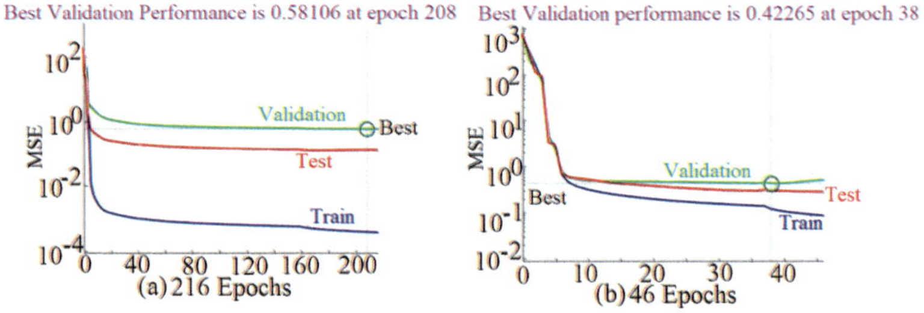

Fig. 5.16 Performance of ANN model: **a** first stage, **b** second stage

(6.4%) are realized, respectively with a peak broadside gain of about 7.36 dBi. The fabricated antenna prototype is shown in Fig. 5.17b.

5.2.5 Half U-Slot Cut SMSA for CP Response

As discussed earlier, orthogonal polarization, dual polarization, and CP response are required so that MSA can receive signals from any polarization of the incident wave. Circular polarization can be achieved by creating asymmetry in the MSA structure. Based on this concept, offset E-shape MSA, and half E-shape MSA discussed in [9], and asymmetric U-slot cut MSA reported in [10], realizes CP response while exciting the orthogonal modes. A single and double U-slot cut MSA discussed in [8] realizes increased cross-polarization radiation pattern, due to the modified higher order resonant modes present. In continuation with [8], a compact half U-slot cut

Table 5.10 Two-step ANN optimized practical dimensions

F	h	x_f	h_a	l_s	w_s	y	R
900	0.07λ	− 3.3	2.2	9.1	0.8	2.4	8.5
900	0.08λ	− 3.5	2.6	9.6	0.8	2.3	8.2
1500	0.06λ	− 1.2	1	3.6	0.6	1.2	4.9
1500	0.07λ	− 1.8	1.4	5.4	0.6	1.2	4.8
1500	0.08λ	− 1.9	1.5	5.7	0.6	1.2	4.6
2700	0.08λ	− 1.1	0.7	2.9	0.4	0.7	2.5
3300	0.06λ	− 0.5	0.6	1.6	0.3	0.7	2.1
3300	0.07λ	− 0.5	0.6	1.6	0.3	0.7	2.1
3300	0.08λ	− 0.5	0.6	1.6	0.3	0.7	2.1
3900	0.06λ	− 0.4	0.5	1.5	0.2	0.6	1.9
3900	0.08λ	− 0.5	0.5	1.4	0.2	0.5	1.9
4300	0.06λ	− 0.4	0.5	1.4	0.2	0.5	1.8
4300	0.08λ	− 0.4	0.5	1.4	0.2	0.5	1.8
5100	0.07λ	− 0.3	0.5	1.3	0.2	0.5	1.6
5100	0.08λ	− 0.3	0.5	1.3	0.2	0.5	1.6
5500	0.06λ	− 0.3	0.4	1.2	0.2	0.5	1.6
5500	0.07λ	− 0.3	0.4	1.2	0.2	0.5	1.6

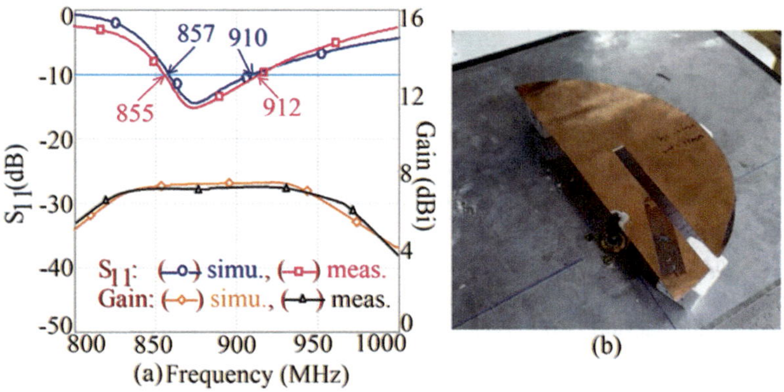

Fig. 5.17 **a** Simulated and measured S_{11} BW and gain plots, and **b** fabricated prototype of rectangular slot cut SCMSA designed at 900 MHz on 0.07λ thickness

square MSA for dual band characteristics with a CP response in the second band is reported in [11]. This configuration of half U-slot cut SMSA ($L = W$) is shown in Fig. 5.18a, b. It was explained in [11] that due to equal values of SMSA dimensions along two axes, TM_{10} and TM_{01} modes are excited at the same frequency. With the offset feed point location, TM_{11} mode is also present. The separation of TM_{10} and TM_{01} modes takes place with the introduction of vertical slot L_{v1}. With an increment

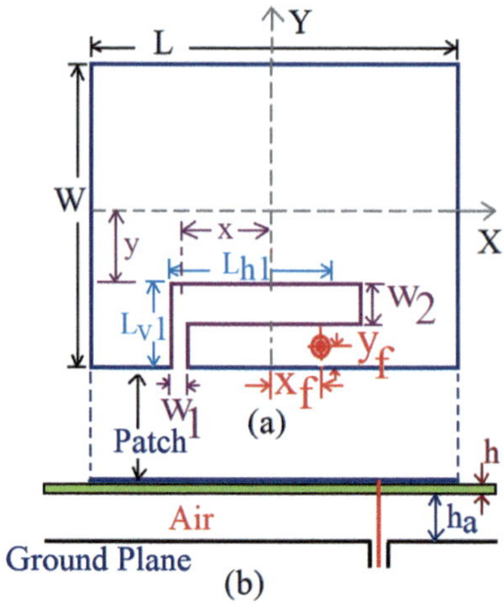

Fig. 5.18 a, b Dual band half U-slot cut square MSA with CP response in the second frequency band [11]

in L_{v1}, reduction in TM_{10} and TM_{11} mode frequencies is observed. This is due to the orthogonal orientation of L_{v1} with respect to the TM_{10} and TM_{11} modal currents.

However, smaller reduction in TM_{10} mode frequency is noted as vertical half U-slot length position is not at the maximum current location of TM_{10} mode [11]. Further with the introduction of L_{h1} and its length increment, TM_{10} and TM_{11} mode frequencies start reducing, whereas TM_{01} mode frequency remains constant. The surface current distributions across TM_{10} and TM_{01} modes are orthogonal in nature. With the increment in length L_{h1}, the impedance at TM_{01} and TM_{11} modes decrease, whereas that at TM_{10} mode increases. Also, an increase in the horizontal half U-slot length yields lower AR value around the modified TM_{11} mode frequency, leading to the possible CP response [11]. However, with the separation in between TM_{10} and TM_{01} resonant mode frequencies, AR value increases in the lower frequency band, i.e., nearer to the modified TM_{01} mode frequency [11]. In the lower frequency band around TM_{01} mode frequency, dual polarized response is realized due to the orthogonal nature of surface currents of TM_{10} and TM_{01} modes. With the orthogonal current distribution at modified TM_{11} mode along square patch length and width, CP is realized in the second band. The results of half U-slot MSA as reported in [11] are shown in Fig. 5.19.

Further, the ANN model for half U-slot cut square MSA for dual polarized response in the first of the dual band and circular polarized response in the second band is proposed. Initially, ANN training data set is created using the following procedure. The detailed procedure of the training data set creation is given in the flowchart as shown in Fig. 5.20.

1. Design SMSA at a given TM_{10} mode resonance frequency f_r and substrate thickness h_t using resonance frequency equation.
2. Initial value of dielectric constant is chosen as $\varepsilon_{re} = 1.06$, as suspended dielectric configuration is considered for the design of half U-slot cut SMSA.

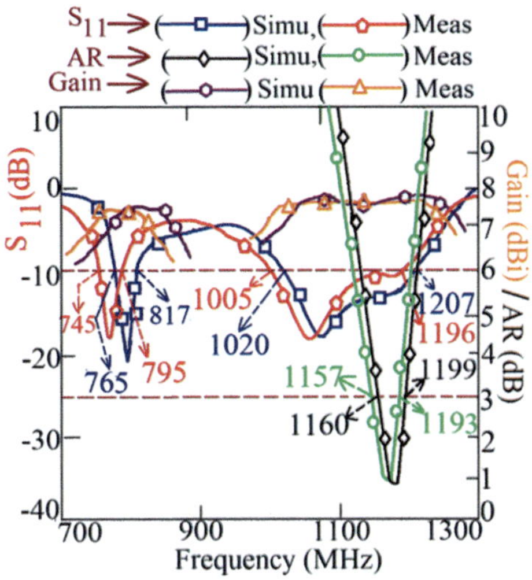

Fig. 5.19 S_{11}, AR BW, and broadside gain plots for dual band half U-slot cut SMSA with CP response in the second band [11]

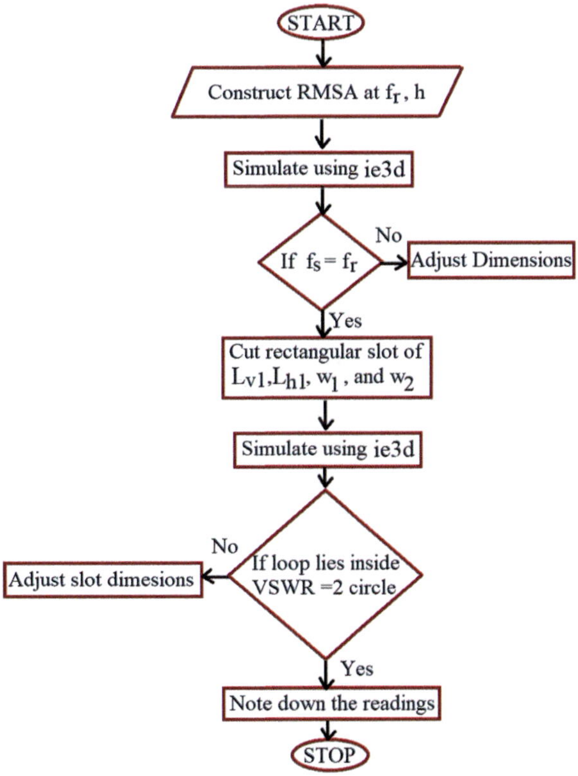

Fig. 5.20 Flowchart of training data set creation of half U-slot cut SMSA

3. If the simulated frequency observed is not as per desired value, adjust the patch dimensions.
4. Simulate the configuration for half U-slot dimensions to achieve the dual band response with CP characteristics in the second frequency band.
5. Optimize the configuration employing parametric process for half U-slot dimension till the optimum response is achieved.
6. Note down the antenna parameters for the dual band response.

The above procedure is repeated for a frequency range of 800–3000 MHz in steps of 200 MHz and substrate thicknesses ranging from $0.06\lambda_g$ to $0.08\lambda_g$. Here the air-suspended FR4 substrate is selected, and thus guided wavelength (λ_g) is considered for the substrate thickness. Sample training data set created using the above procedure is given in Table 5.11.

Table 5.11 Sample ANN training data set of half U-slot MSA for DP and CP response

f_r	h_t (in λ_g)	h + ha	L	w_1	w_2	x_1	y_1	x_f	y_f
700	0.06	3.122	20	3.7	1.5	3.8	4	2.6	1.7
900	0.06	2.428	15.5	2.87	1.181	3	3.026	2.2	1
1100	0.06	1.987	12.8	2.4	1	2.4	2.4	1.7	0.9
1300	0.06	1.681	10.8	1.989	0.819	2.1	2.124	1.3	0.7
1500	0.06	1.457	9.3	1.7	0.7	1.8	1.85	1.25	0.65
1700	0.06	1.285	8.2	1.533	0.627	1.6	1.592	1.2	0.55
1900	0.06	1.15	7.4	1.363	0.561	1.4	1.456	1	0.5
2100	0.06	1.04	6.7	1.2	0.5	1.3	1.35	0.89	0.6
2300	0.06	0.95	6.1	1.175	0.5	1.2	1.05	0.9	0.6
2500	0.06	0.874	5.6	1.09	0.482	1.1	0.87	0.85	0.55
2700	0.06	0.809	5.2	1	0.4	1	1	0.77	0.55
2900	0.06	0.754	4.8	0.9	0.485	0.9	0.46	0.4	0.65
700	0.08	4.16	18	3.3	1.4	3.4	3.4	1.6	0.6
900	0.08	3.237	14	2.6	1.1	2.7	2.6	1.3	0.5
1100	0.08	2.649	11.4	2	1	2.2	3.5	1.6	0.7
1300	0.08	2.2	9.6	1.8	0.75	1.9	1.8	1.2	0.8
1500	0.08	1.9	8.3	1.425	0.662	1.6	1.5	1.1	0.75
1700	0.08	1.7	7.4	0.662	0.55	1.4	1.5	0.8	0.7
1900	0.08	1.5	6.6	1.2	0.575	1.3	1	0.745	0.6
2100	0.08	1.4	6	1.1	0.525	1.2	0.9	0.625	0.5
2300	0.08	1.3	5.5	1	0.45	1.1	0.95	0.55	0.355
2500	0.08	1.2	5	0.9	0.475	1	0.6	0.445	0.605
2700	0.08	1.1	4.7	0.9	0.45	0.9	0.55	0.5	0.45
2900	0.08	1	4.3	0.8	0.45	0.8	0.35	0.345	0.5

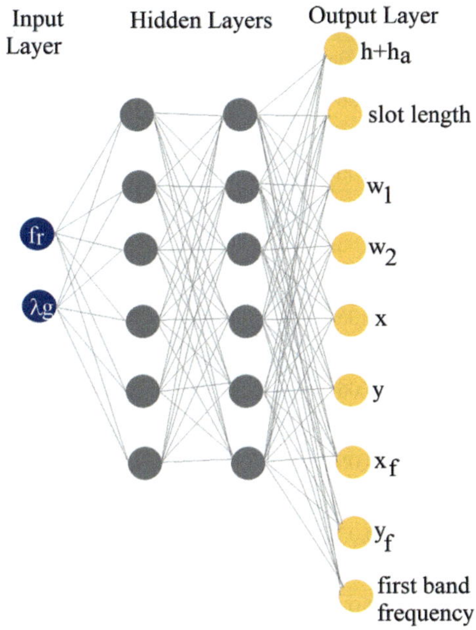

Fig. 5.21 ANN configuration of half U-slot cut MSA for DP and CP in second band

A detailed structure of ANN model for the trained data is shown in Fig. 5.21. A four-layer network is used in the ANN model of half U-slot cut SMSA. The first layer is the input layer, with 2 neurons having frequency and wavelength (λ_g) as features.

There are two hidden layers, one with 6 neurons with ReLU activation function to provide nonlinearity to the model and the other with all 9 neurons as outputs for half U-slot cut SMSA parameters with sigmoid activation function. To prevent a slow learning gradient descent curve for the training model, MinMaxScalar is used to scale the features. Because the generated data set has no outliers, mean squared error is utilized as the loss function. Adam optimizer is used in the ANN because of its twin benefits of stochastic gradient descent with momentum and adaptive learning of learning rate. The detailed table of the ANN architecture with number of inputs, outputs, and hidden layers is like Table. 2.6, except for the number of inputs. Here in this case, they are 9. Using the proposed ANN model, which got trained with the above training data set, antenna dimensions are predicted for a frequency range of 800–3000 MHz in steps of 200 MHz. The dimensions predicted by the ANN architecture are given in Table 5.12, after minor changes. These minor changes in the dimensions are done so as to obtain practically realizable values as shown in Table 5.12. The mean squared errors of trained and predicted dimensions by ANN with varying epochs are shown in Fig. 5.22. A reduction in error is observed with a large number of epochs.

From the dimensions observed from the table, the half U-slot SMSA is designed at 1200 MHz for $h_t = 0.08\lambda_g$, and simulations are conducted to verify the result in terms of dual band and CP response. All the simulated and measured results are

Table 5.12 ANN predicted dimensions of half U-slot cut MSA for DP and CP response

f_r	h_t (in λ_g)	L	w_1	w_2	x_1	y_1	x_f	y_f
800	0.06	16.817	3.65	1.252	3.012	6.073	2.4	1
1000	0.06	14.583	2.64	1.129	2.81	4.837	2	0.9
1200	0.06	12.445	2.284	0.903	2.398	4.439	1.7	0.8
1400	0.06	10.379	1.931	0.755	1.996	3.039	1.5	0.7
1600	0.06	8.995	1.694	0.645	1.717	2.761	1.3	0.6
1800	0.06	7.861	1.481	0.572	1.495	2.511	1.1	0.5
2000	0.06	6.99	1.298	0.528	1.331	2.299	1	0.5
2200	0.06	6.355	1.147	0.504	1.216	3.125	0.9	0.5
2400	0.06	5.909	1.026	0.49	1.138	1.988	0.9	0.4
2600	0.06	5.603	0.933	0.483	1.088	1.884	0.8	0.4
2800	0.06	5.397	0.861	0.409	1.055	1.806	0.7	0.4
3000	0.06	5.26	0.808	0.477	1.035	2.748	0.7	0.3
800	0.08	14.994	2.708	1.26	2.945	4.852	1.3	1
1000	0.08	13.097	2.407	1.021	2.591	2.486	1.2	0.9
1200	0.08	10.927	2.039	0.785	2.17	2.171	1.2	0.8
1400	0.08	9.002	1.686	0.705	1.788	1.682	0.8	0.7
1600	0.08	7.528	1.385	0.619	1.493	2.356	0.7	0.6
1800	0.08	6.522	1.151	0.546	1.292	2.106	0.6	0.5
2000	0.08	5.887	0.982	0.509	1.167	0.929	0.46	0.45
2200	0.08	5.507	0.867	0.431	1.093	0.809	0.42	0.4
2400	0.08	5.323	0.807	0.483	1.057	0.746	0.33	0.3
2600	0.08	5.211	0.768	0.379	1.036	0.704	0.3	0.2
2800	0.08	5.138	0.739	0.477	1.022	0.674	0.27	0.125
3000	0.08	5.09	0.717	0.376	1.014	0.653	0.27	0.1

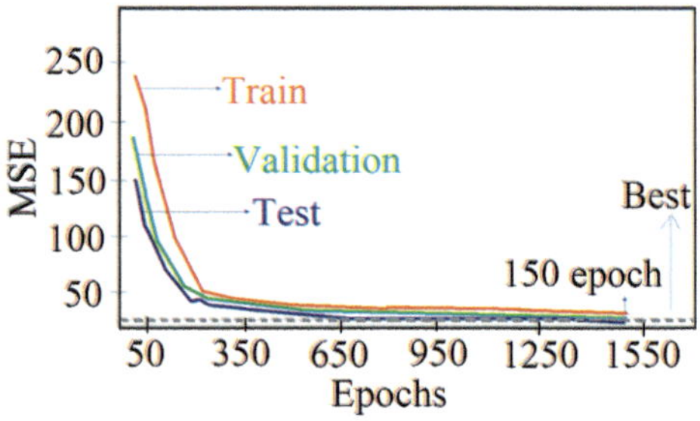

Fig. 5.22 MSE versus Epochs graph

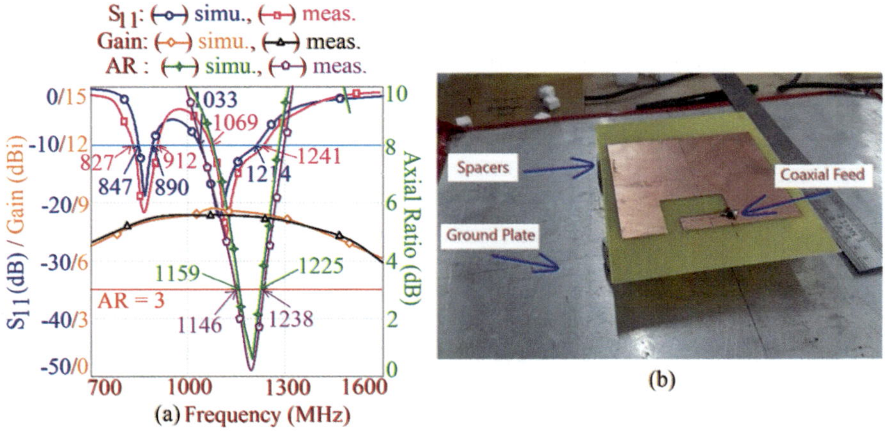

Fig. 5.23 **a** S_{11}, AR BW, and gain plots and **b** fabricated prototype of half U-slot cut SMSA designed from predicted dimensions at 1000 MHz on $0.065\lambda g$ substrate thickness

shown in Fig. 5.23a, and the fabricated prototype of half U-slot cut SMSA designed at 1200 MHz from ANN predicted dimensions is shown in Fig. 5.23b.

The simulated values of the S_{11} BW in the two bands are 43 MHz (4.9%) and 181 MHz (16.1%), respectively. Respective measured values are 85 MHz (9.8%) and 172 MHz (14.89%). The simulated and measured AR BW in the second band is 66 MHz (5.5%) and 92 MHz (7.7%), respectively. In the two bands, broadside peak gain is about 8.67 dBi and offers stable gain characteristics across the impedance and AR BW. In both the bands, the antenna exhibits a broadside radiation pattern. In the second band, due to the CP response, radiation pattern plot shows equal contribution of co- and cross-polar levels.

5.3 Summary

The ANN modelling of compact variations of E-shape and U-slot cut MSAs for broadband and CP responses is presented. For each MSA its functioning is explained with respect to its resonance mode behaviour, and then ANN model details along with nodal details, algorithm used, and criterion of the ANN such as MSE are provided. All the models utilized the double step ANN process. The ANN with a double-stage approach predicts dimensions of MSAs, which yields better results in terms of performance curve and regression graphs. In modelling of all compact MSAs, ReLU activation function and MinMaxScalar are used to scale the features. With the help of the ANN model, the dimensions of compact E-shape and U-slot cut MSAs can be predicted at a given frequency and substrate thickness, which helps in reducing the time required to optimize MSA dimensions while simulating them individually. A slight modification in the ANN predicted dimensions may be required to round off the dimensions to the nearest integer to get practically realizable dimensions. In all the MSAs, ANN predicted dimensions and calculated dimensions shown less than 5% variation.

References

1. Wong KL (1999) Design of nonplanar microstrip antennas and transmission lines. Wiley, New York
2. Kumar G, Ray KP (2003) Broadband microstrip antennas. Artech House, Boston
3. Sanad M (1994) Effect of shorting post on short circuit microstrip antennas. In: Proceedings of IEEE-APS, Seattle, WA, pp 794–797
4. Lee KF, Huynh T (1995) Single-layer single-patch wideband microstrip antenna. Electron Lett 31(16):1310–1312. https://doi.org/10.1049/el:19950950
5. Deshmukh AA, Kumar G (2005) Compact broadband E-shaped microstrip antennas. Electron Lett 41(18):1
6. Deshmukh AA, Kumar G (2005) Compact broadband U-slot-loaded rectangular microstrip antennas. Microw Opt Technol Lett 46(6):556–559
7. Chavali VAP, Deshmukh AA (2020) Half U-slot and rectangular slot loaded nearly square microstrip antennas for wideband response. Prog Electromagn Res C 104:129–141. https://doi.org/10.2528/pierc20060205
8. Deshmukh AA, Jain AR, Joshi AA (2013) Analysis of broadband slot cut semi-circular microstrip antennas. Int J Comput Appl 975:8887
9. Chavali VA, Deshmukh AA (2022) Wideband and circularly polarized designs of modified E-shape microstrip antennas for GSM and GPS applications. Prog Electromagn Res C 121
10. Lee KF, Luk KM, Mok WC, Nayeri P (2011) Single probe-fed circularly polarized patch antennas with U-slots. Microw Opt Technol Lett 53(6):1245–1253
11. Ambekar AG, Deshmukh AA (2022) Dual band compact square microstrip antenna for GSM and GPS applications. Prog Electromagn Res C 118

GPSR Compliance

The European Union's (EU) General Product Safety Regulation (GPSR) is a set of rules that requires consumer products to be safe and our obligations to ensure this.

If you have any concerns about our products, you can contact us on ProductSafety@springernature.com

In case Publisher is established outside the EU, the EU authorized representative is:

Springer Nature Customer Service Center GmbH
Europaplatz 3
69115 Heidelberg, Germany

Batch number: 10704307

Printed by Printforce, the Netherlands